AF428570

The Space Race

TD Barnes

Published by TD Barnes, 2024.

While every precaution has been taken in the preparation of this book, the publisher assumes no responsibility for errors or omissions, or for damages resulting from the use of the information contained herein.

THE SPACE RACE

First edition. October 16, 2024.

Copyright © 2024 TD Barnes.

ISBN: 979-8224725212

Written by TD Barnes.

Table of Contents

The Space Race

By TD Barnes

Copyright 2024 TD Barnes

Foreword

The Space Race was one of the defining struggles of the 20th century, a clash not only between two superpowers but also between two ideologies—each vying for supremacy in space. What began as a battle for geopolitical dominance soon turned into a remarkable journey of human discovery, technological innovation, and exploration. The vast expanse beyond Earth's atmosphere became the new frontier, with both the United States and the Soviet Union racing to claim it.

In the years that followed the launch of Sputnik 1 in 1957, both nations made breathtaking advances that would forever alter humanity's place in the cosmos. Rockets capable of carrying satellites, animals, and, eventually, humans into space were developed with unprecedented speed. This era of rapid scientific progress, driven by political urgency, gave rise to some of the most iconic achievements in history: the first artificial satellite, the first human spaceflight, the first landing on the Moon.

But this story is not merely one of competition; it is also one of cooperation and the gradual realization that space, though contested during the Cold War, belongs to all of humanity. The Cold War rivalry eventually paved the way for collaboration between former adversaries, as evidenced by the Apollo-Soyuz Test Project in 1975 and the creation of the International Space Station decades later.

This book tells the story of that race to the stars—a race that shaped the modern world and continues to influence space exploration today. The journeys chronicled in these pages reveal the triumphs and tragedies, the scientific breakthroughs, and the spirit of adventure that drove humanity to the edge of space and beyond.

As we look to the future, with private companies and new nations entering the space exploration arena, the Space Race legacy remains ever-present. It was a race that sparked the imagination of millions and laid the foundation for a future where space is not just a battleground for superpowers but a shared frontier for all of humanity.

Chapter 1 - The Cold War

The first nuclear weapon, developed by the United States during World War II, was a direct response to the fear that the Axis powers, particularly Nazi Germany, might develop such a weapon first. This project, known as the Manhattan Project, was an unprecedented scientific and military effort that brought together some of the world's leading scientists to develop nuclear technology. The goal was to create a bomb capable of ending the war swiftly and decisively.

While the U.S. was advancing in nuclear weapons research, the Soviet Union, under Joseph Stalin, had also been conducting research in nuclear physics. Soviet scientists recognized the destructive potential of nuclear weapons, but their program lagged behind that of the U.S. The Soviet Union was not officially informed of the Manhattan Project, the secretive U.S. effort to develop the bomb, despite being an ally during the war. It wasn't until the Potsdam Conference in July 1945, shortly after the successful test of the first atomic bomb in New Mexico, that U.S. President Harry S. Truman informed Stalin of the weapon's existence. Truman expected a surprised reaction from Stalin, but Stalin remained calm—likely because Soviet intelligence had already learned about the project through spies working within the Manhattan Project itself.

Soviet spies, including Klaus Fuchs and Theodore Hall, provided Stalin with detailed information about the bomb's development. Despite the tight security surrounding the Manhattan Project, these spies succeeded in delivering critical designs to the Soviets, allowing them to make rapid advancements in their own nuclear program. This espionage network would eventually be uncovered, leading to the arrest of several individuals involved in the espionage, such as Fuchs, Harry Gold, and Julius and Ethel Rosenberg, the latter two being executed for their role in passing atomic secrets to the Soviets.

The U.S. dropped two atomic bombs on Japan in August 1945—one on Hiroshima and the other on Nagasaki—effectively ending World War II. The bombings marked the dawn of the nuclear age, with the United States initially possessing a monopoly on nuclear weapons. However, the Soviet Union was not far behind, thanks in part to the intelligence gathered from its spy network.

The Cold War emerged after World War II as geopolitical tensions rose between the two leading superpowers: the United States and the Soviet Union. These two nations, which had been wartime allies against Nazi Germany, soon found themselves at odds due to their differing political ideologies and visions for the postwar world.

The United States championed democracy and capitalism, promoting free markets and individual liberties, while the Soviet Union sought to spread communism, advocating a state-controlled economy and a single-party system. These ideological differences created a deep mistrust between the two nations.

Several key events fueled the rise of the Cold War. At the Yalta and Potsdam Conferences in 1945, the Allied powers attempted to negotiate the future of postwar Europe. Disagreements arose over how to manage war-torn Germany, the fate of Eastern European nations, and the spread of communism. The Soviet Union sought to expand its influence over Eastern Europe, while the U.S. and its allies wanted to ensure democratic governance.

After the war, the Soviet Union established communist governments across Eastern Europe, effectively creating a buffer zone between itself and Western Europe. This expansion alarmed the West, which viewed it as an attempt to spread communism.

After World War II, several countries in Eastern Europe were not directly annexed into the Soviet Union but were turned into Soviet satellite states. These satellite states were countries that, while technically independent, were heavily influenced and controlled by the Soviet Union, especially in terms of political ideology, governance, and military alignment. The Soviet Union's occupation and influence in these countries were part of its broader strategy to create a buffer zone between itself and the capitalist West, ensuring its security and the spread of communism.

Following Germany's division after World War II, the country's eastern portion became a Soviet-controlled satellite state. East Germany was established in 1949 and was closely aligned with the Soviet Union, both politically and militarily. The Berlin Wall, built in 1961, became a stark symbol of the Cold War divide between East and West Germany.

The People's Republic of Poland, which had suffered greatly during the war, came under Soviet influence after the Red Army liberated it from Nazi control. A communist government was established, and Poland became a satellite state, with Soviet troops stationed on its soil and its political system aligned with Moscow.

After the war, the People's Republic of Hungary fell under Soviet control, and a communist regime was installed in 1949. Hungary attempted to break away from Soviet dominance during the Hungarian Revolution of 1956, but Soviet forces crushed the uprising, reaffirming Hungary's status as a satellite state.

The Czechoslovak Socialist Republic initially enjoyed some degree of political independence after the war, but by 1948, a communist coup d'état brought the country firmly under Soviet influence. Like Hungary, Czechoslovakia experienced a brief period of reform and resistance in 1968 (the Prague Spring), but this was also suppressed by Soviet military intervention.

The People's Republic of Romania became a Soviet satellite state after World War II, with a communist government installed in 1947. While initially loyal to the Soviet Union, Romania's leadership, under Nicolae Ceaușescu, pursued a more independent foreign policy in later years, although it remained a communist dictatorship.

The People's Republic of Albania became a satellite state of the Soviet Union after the war, but by the 1960s, it aligned itself more closely with the People's Republic of China. Albania's leader, Enver Hoxha, sought to distance the country from Soviet influence after the Sino-Soviet split, making Albania unique among the Eastern Bloc countries.

These satellite states were key components of the Eastern Bloc and the Warsaw Pact, a military alliance formed in 1955 as a counterbalance to NATO. The Soviet Union maintained strict control over these countries through their communist governments, secret police, and the presence of Soviet military forces. While some of these states tried to break free from Soviet dominance at various points during the Cold War, the Soviet Union responded with force, reaffirming its grip on Eastern Europe until the eventual collapse of communism in the region in the late 1980s.

The Iron Curtain, coined by Winston Churchill in a 1946 speech, symbolized the growing division between the communist East and the capitalist West. It reflected the physical and ideological barrier that separated Soviet-controlled territories from the rest of Europe.

In 1947, U.S. President Harry Truman introduced the Truman Doctrine, which promised support for countries resisting communism. This policy was part of a broader strategy of containment aimed at preventing the spread of communism beyond where it already existed.

The following year, the U.S. launched the Marshall Plan to provide economic aid to rebuild war-torn Europe. The Soviet Union saw this as an attempt by the West to undermine communist influence, deepening the divide. This spurned the Berlin Blockade and Airlift (1948-1949), one of the first major confrontations of the Cold War, when the Soviet Union blocked access to West Berlin to force the Western Allies to abandon the city. The U.S. and its allies responded by airlifting supplies to the besieged city, demonstrating their commitment to resisting Soviet pressure.

Following the war, the international community, recognizing the catastrophic potential of nuclear weapons, sought ways to prevent their proliferation. The United Nations, newly founded in 1945, convened its first General Assembly in London in 1946. One of the primary topics discussed was the future of nuclear weapons, which led to the creation of the United Nations Atomic Energy Commission. The U.S. presented the Baruch Plan, which called for an international authority to oversee atomic energy and prevent the spread of nuclear weapons. However, the Soviet Union rejected this proposal, instead advocating for universal nuclear disarmament. Both proposals were ultimately rejected, signaling the beginning of the nuclear arms race that would define much of the Cold War era.

The first nuclear weapon, developed by the United States during World War II, was created in response to concerns that Nazi Germany might develop such a weapon first. This project, known as the Manhattan Project, was an unprecedented scientific and military initiative that brought together some of the world's leading scientists to harness nuclear energy for warfare. The primary objective was to create a bomb capable of ending the war quickly and decisively, altering the course of global conflict.

While the U.S. progressed in its nuclear research, the Soviet Union, under Joseph Stalin, had also been conducting studies in nuclear physics. Soviet scientists recognized the destructive potential of nuclear weapons but were far behind the U.S. in development. Although the U.S. and the Soviet Union were allies during the war, the Soviets were not officially informed about the Manhattan Project. It was only at the Potsdam Conference in July 1945, after the successful test of the atomic bomb in New Mexico, that U.S. President Harry S. Truman disclosed its existence to Stalin. Truman expected a dramatic reaction from Stalin, but the Soviet leader remained unusually calm, likely because Soviet intelligence, through espionage, was already well-informed about the project.

Spies like Klaus Fuchs and Theodore Hall had infiltrated the Manhattan Project and passed vital information to the Soviet Union. These spies provided Stalin with critical details on the bomb's design, including insights into both the implosion bomb and later the hydrogen bomb. Despite the Manhattan Project's extreme security measures, this espionage network allowed the Soviet Union to advance its own nuclear program rapidly. Once exposed, the espionage network was dismantled, and key figures, including Fuchs and American citizens Julius and Ethel Rosenberg, were arrested. The Rosenbergs were later executed for their roles in passing atomic secrets to the Soviets.

In August 1945, under orders from President Truman, the U.S. dropped two atomic bombs—one on Hiroshima and the other on Nagasaki—forcing Japan's surrender and effectively ending World War II. These bombings marked the beginning of the nuclear age, with the U.S. initially holding a monopoly on nuclear weapons. However, the Soviet Union, thanks in part to intelligence gathered through espionage, was soon able to develop its own nuclear arsenal, contributing to the intense rivalry that would shape the Cold War.

In the postwar years, the international community, recognizing the catastrophic potential of nuclear weapons, began discussing ways to control and prevent their proliferation. The United Nations, founded in 1945, convened its first General Assembly in 1946, during which the future of nuclear weapons was a key topic. This led to the creation of the United Nations Atomic Energy Commission, which aimed to control the use of nuclear energy. The U.S. presented the Baruch Plan, which proposed placing all atomic activities under international control to prevent further nuclear weapons development. The Soviet Union, however, rejected this idea and instead proposed universal nuclear disarmament. Both proposals were ultimately rejected, setting the stage for the nuclear arms race that would dominate global politics during the Cold War.

The Soviet Union's development of rocket technology traces its origins back to the early 20th century, beginning with experimental efforts that laid the groundwork for the remarkable achievements of the Space Race (Russian: Космическая гонка [kɐsˈmʲitɕɪskəjə ˈɡonkə]). In 1921, the Soviet military established the Gas Dynamics Laboratory to investigate the potential of solid-fuel rockets. Led by Nikolai Tikhomirov, who had been exploring rocket technology since the late 19th century, the laboratory's research would become foundational. Tikhomirov, a pioneer in the field, had already patented a design for self-propelled aerial and water-surface mines in 1915. By 1928, the Soviet Union had successfully carried out its first test-firing of a solid-fuel rocket.

The 1930s marked significant advances in Soviet rocketry, with the formation of the Group for the Study of Reactive Motion (GIRD). This group, led by Sergey Korolev, Friedrich Zander, Mikhail Tikhonravov, and Leonid Dushkin, conducted crucial research that culminated in the launch of the GIRD-X in 1933, the first Soviet liquid-fueled rocket. That same year, the consolidation of Soviet rocket research entities resulted in the creation of the Reactive Scientific Research Institute. This institute developed several key rocket designs, including the RP-318, the Soviet Union's first rocket-powered aircraft, and the RS-82 and RS-132 missiles, which later served as the basis for the famous Katyusha multiple rocket launcher.

Although Soviet rocket technology was advancing rapidly, Joseph Stalin's Great Purge from 1936 to 1938 significantly disrupted this progress. Key engineers and scientists were arrested or executed, setting back the country's development efforts. However, by the end of World War II, the Soviet Union received a significant boost when it captured Nazi Germany's A-4 (V-2) rocket production facilities, along with several key German scientists and engineers. This acquisition enabled Soviet scientists to reverse-engineer the A-4 and develop their own version, the R-1, under the leadership of Sergey Korolev. The R-1 became operational in the Soviet Army by November 1950.

Building on the experience gained from the R-1, Korolev, along with Valentin Glushko, designed the R-2 rocket, which had an improved range of 600 kilometers (370 miles) and entered service in 1951. The next major breakthrough came in 1955 with the introduction of the R-5 Pobeda, the Soviet Union's first true strategic missile, capable of carrying a 1 megaton thermonuclear warhead over 1,200 kilometers (750 miles). These rockets, along with scientific versions of the R-1, R-2, and R-5, were used for a variety of experiments, including sending space dogs into suborbital flights to study the effects of space travel on living organisms.

In 1953, design work began on a much more ambitious project: the R-7 Semyorka. This intercontinental ballistic missile was designed to launch a 3,000-kilogram nuclear warhead over a range of 8,500 kilometers, enough to strike the United States. As development progressed, the warhead mass was increased to accommodate advancements in thermonuclear bomb technology. On August 21, 1957, the R-7 successfully flew 6,000 kilometers (3,700 miles), becoming the world's first intercontinental ballistic missile.

Just two months later, the R-7 achieved another historic milestone when it launched Sputnik 1, the world's first artificial satellite, into orbit on October 4, 1957. This launch ushered in the Space Age and firmly established the Soviet Union as a formidable player in space exploration. The R-7 would go on to serve as the foundation for a family of space launch vehicles, including those used for the Sputnik, Vostok, Voskhod, and Soyuz programs, as well as the Molniya and Luna missions. Variants of the R-7 remain in use today, making it one of the most reliable and enduring rocket systems in space history.

Through these early efforts in rocketry, the Soviet Union laid the groundwork for its later achievements in space exploration, from launching the first satellite to sending the first human into space. The innovations of the R-7 and its successors would go on to play a crucial role in the intense competition between the Soviet Union and the United States during the Space Race.

American rocket development during the mid-20th century was characterized by the integration of pioneering work from both domestic scientists and expatriate engineers, most notably Wernher von Braun. Although American rocket engineer Robert H. Goddard had laid the foundations for liquid-propellant rocket technology as early as 1914, his groundbreaking ideas were largely dismissed, particularly after a critical editorial in *The New York Times*. This public ridicule drove Goddard into relative isolation, and as a result, the United States lagged behind the other major World War II powers—Germany and the Soviet Union—in rocket development.

By the end of World War II, however, the U.S. gained a significant advantage in rocketry through Operation Paperclip, a secret program to recruit Nazi German scientists, engineers, and technicians. Among the most prominent of these recruits was Wernher von Braun, the architect of Germany's V-2 rocket program. In 1945, von Braun and most of his engineering team were brought to the United States, along with a substantial number of captured V-2 rockets. Initially stationed at the Army's White Sands Proving Ground in New Mexico, von Braun's team began assembling and launching the V-2 rockets, conducting vital tests that would inform the future of American rocket technology. These early tests yielded the first photographs of Earth from space and led to the creation of the WAC Corporal-V-2, a two-stage rocket successfully launched in 1949.

In 1950, the German rocket team was relocated to the Army's Redstone Arsenal in Huntsville, Alabama, marking a turning point in U.S. rocket development. From this new base, von Braun and his engineers developed the Army's first operational medium-range ballistic missile, the Redstone rocket. The Redstone would later serve as the launch vehicle for America's first satellite, Explorer 1, and the first crewed space missions under NASA's Project Mercury. This rocket became the foundation for further American missile systems, including the Jupiter and Saturn families of rockets, the latter of which would eventually carry astronauts to the Moon.

By the mid-1950s, rocket development in the U.S. expanded across multiple branches of the military, each developing its own ballistic missile programs. The Air Force had begun its own intercontinental ballistic missile (ICBM) research in 1945 with the MX-774 project, an initiative that laid the groundwork for future rocket designs. In parallel, von Braun's team was conducting tests on the Air Force's PGM-11 Redstone rocket at Cape Canaveral, Florida. These tests demonstrated the increasing sophistication of American rocket technology and set the stage for future space exploration efforts.

A significant milestone in U.S. rocketry came in 1957, when the Air Force's ICBM program culminated in the Atlas-A, the first successful American ICBM. The Atlas program would soon evolve into the Atlas-D, which not only served as a nuclear-capable ICBM but also played a crucial role in space exploration. The Atlas-D launched the first American astronauts during Project Mercury and was used as the launch vehicle for the Agena Target Vehicle, an essential component of the orbital rendezvous techniques developed during Project Gemini.

As the U.S. solidified its leadership in space exploration, the development of increasingly powerful rockets, such as the Saturn IB and later the Saturn V, allowed NASA to embark on ambitious missions, culminating in the Apollo lunar landings. These efforts were supported by a robust stable of launch vehicles, including those adapted from ICBMs, such as the Atlas and Titan series, as well as purpose-built rockets like the Saturn family. Together, these advancements laid the foundation for America's space program, enabling it to achieve historic milestones in human spaceflight and scientific exploration.

Through these combined efforts—spearheaded by both American pioneers and expatriate experts—the United States transitioned from a relative laggard in rocket technology to a leader in space exploration, a status that would define the nation's role in the Space Race and beyond.

On August 30, 1955, Sergei Korolev, the chief architect of the Soviet space program, successfully convinced the Soviet Academy of Sciences to establish a commission dedicated to beating the United States into Earth's orbit. This marked the unofficial beginning of the Space Race. The Soviet Union's Council of Ministers quickly classified the development of the space program as top-secret, initiating a policy of concealment that would characterize the Soviet approach for decades. The decision to launch Sputnik, the world's first artificial satellite, was taken under the strictest secrecy, and the Politburo carefully considered how to present this monumental achievement to the world.

When Sputnik's launch was approved, the Telegraph Agency of the Soviet Union (TASS) was tasked with crafting all official announcements. True to Soviet policy, the information released after Sputnik's successful launch on October 4, 1957, offered little insight into the individuals or institutions responsible for the achievement. No mention was made of Korolev or the team of scientists and engineers who made history. Instead, the announcement was laden with complex scientific jargon, designed to obscure as much as it revealed. One historian noted that the public statement contained "an abundance of arcane scientific and technical data... as if to overwhelm the reader with mathematics in the absence of even a picture of the object." This calculated ambiguity heightened the sense of mystery surrounding the Soviet space program, adding to its enigma on the global stage.

Secrecy became a hallmark of the Soviet space effort, serving both to protect classified information from the West and to keep the Soviet populace in the dark. The Soviet government maintained a controlled narrative, releasing little more than triumphant headlines, while details of missions were often withheld or distorted. Launches were not publicized until they had already occurred, and the names of cosmonauts were only revealed after they were safely in orbit. Critical information about the size, design, and capabilities of Soviet rockets, spacecraft, and their occupants remained classified. Only the early successes, such as the Sputnik satellites, lunar probes, and Venus missions, were celebrated publicly.

The Soviet military exercised tight control over the space program, with Korolev's design bureau, OKB-1, placed under the Ministry of General Machine Building, an organization primarily focused on developing intercontinental ballistic missiles. This connection between the space program and military technology reinforced the secrecy, as the space program's resources and developments were considered critical to national security. Even into the 1960s, Soviet space projects were often given random numerical identifiers to obscure their purpose and progress.

Public statements were always optimistic, leaving the impression that the Soviet space program was an unbroken series of successes. Failures and setbacks were seldom acknowledged. As historian James Andrews observed, reports on Soviet space missions, especially human spaceflights, rarely mentioned difficulties. The program's shortcomings, such as failed launches or technical problems, were concealed from both the Soviet people and the rest of the world.

This shroud of secrecy extended far beyond the technical details. Dominic Phelan, in *Cold War Space Sleuths*, captured the prevailing mystery of the Soviet space program by quoting Winston Churchill's famous description of the Soviet Union as "a riddle, wrapped in a mystery, inside an enigma." The Soviet space effort, though conducted in the public eye, remained veiled by a figurative "space curtain." Unraveling the truth behind Soviet space activities required significant effort by outside observers, and much about the program's inner workings remained hidden for years.

Despite this veil, the achievements of the Soviet space program during the early years of the Space Race were undeniable. The successful launch of Sputnik in 1957, followed by the first human spaceflight by Yuri Gagarin in 1961, cemented the Soviet Union's early lead in the race to conquer space. Yet, the secrecy that surrounded these accomplishments also shrouded the program's challenges, leaving the world to speculate about the true state of Soviet space exploration during one of the most critical chapters in Cold War history.

In the early stages of the Space Race, President Dwight D. Eisenhower faced a dilemma regarding the implications of launching a satellite into orbit. He was concerned that a satellite passing over a foreign nation at an altitude exceeding 100 kilometers (62 miles) might be perceived as a violation of that nation's airspace, particularly by the Soviet Union. Eisenhower feared that such an action could be framed as an illegal overflight, providing the Soviets with an opportunity to claim a propaganda victory. This concern weighed heavily on the president, especially in the tense political atmosphere of the Cold War.

Eisenhower and his advisors, however, held the view that a nation's sovereignty over its airspace did not extend beyond the Kármán line, the boundary between Earth's atmosphere and outer space, generally set at 100 kilometers. In an effort to solidify this position, they utilized the 1957-58 International Geophysical Year (IGY), a global scientific initiative, to conduct satellite launches and establish this principle in international law. By framing these early space activities as purely scientific endeavors, the United States hoped to avoid accusations of militarism while asserting that outer space was free for exploration by all nations.

Another of Eisenhower's concerns was the possibility of being labeled a "warmonger" if military missiles were used to launch satellites. He wanted to avoid provoking the Soviet Union and feared that such an action might trigger an international incident. To prevent this, Eisenhower opted to use the Naval Research Laboratory's Vanguard rocket, a vehicle designed specifically for research purposes, as the United States' primary launch vehicle. The Vanguard program was intended to demonstrate the peaceful intentions of American space activities, in contrast to the growing perception of military involvement in space exploration.

This decision had significant implications for the development of American space technology. Wernher von Braun, the German rocket scientist leading the U.S. Army's missile program, had developed the powerful Jupiter-C rocket, which was capable of launching a satellite into orbit. However, due to its association with military applications, von Braun's team was prohibited from using the Jupiter-C for orbital missions. Instead, they were restricted to suborbital tests to develop reentry vehicle technology for ballistic missiles.

On September 20, 1956, von Braun's team successfully launched a Jupiter-C rocket, which had the capability to send a satellite into orbit. However, under Eisenhower's directives, the mission was limited to a suborbital flight, testing only the rocket's ability to carry and retrieve a reentry vehicle. While this test demonstrated the technological readiness of von Braun's team to participate in the space race, their efforts to launch a satellite were stalled by the president's cautious approach to space exploration during the early Cold War.

Ultimately, Eisenhower's decision to prioritize the peaceful image of the U.S. space program delayed the nation's first satellite launch, allowing the Soviet Union to claim the first major victory in the Space Race with the successful launch of Sputnik on October 4, 1957. However, his administration's efforts to establish the legal framework for satellite overflights helped shape international space law in the years that followed.

Korolev, upon hearing about von Braun's 1956 Jupiter-C test, mistakenly believed that it had been an unsuccessful satellite mission. This misinterpretation spurred him to accelerate his efforts to launch a Soviet satellite, determined not to fall behind in the emerging Space Race. The Soviet R-7 rocket, which was significantly more powerful than any U.S. launch vehicle at the time, provided Korolev with a critical advantage. He decided to fully exploit this capability by developing a large, ambitious satellite called Object D.

Object D, which was to be the primary Soviet satellite, dwarfed the planned U.S. counterparts. Weighing an impressive 1,400 kilograms (3,100 pounds), it was designed to carry 300 kilograms (660 pounds) of scientific instruments. These instruments would be capable of photographing Earth, measuring radiation levels, and analyzing the planet's magnetic field—an ambitious leap ahead of the modest U.S. satellite proposals. The satellite's name, Object D, was chosen to distinguish it from other R-7 payloads (A, B, V, and G), which were reserved for nuclear warheads.

Despite Korolev's ambitions, progress on Object D was slow, with significant delays in both design and manufacturing. Realizing the satellite would not be ready in time, Korolev sought permission in February 1957 from the Soviet Council of Ministers to develop a simpler satellite, which he called *Prosteishy Sputnik* (PS-1), meaning "Simple Satellite." The council agreed and postponed Object D's launch to April 1958. PS-1, soon to be known as Sputnik 1, was a far more basic craft, a metallic sphere just 58 centimeters (23 inches) in diameter and weighing only 83.8 kilograms (185 pounds). It carried no complex scientific instruments like Object D but had two radio transmitters operating on different frequencies, a system to detect meteoroid impacts on its surface, and the ability to measure the density of Earth's thermosphere.

By the summer of 1957, Korolev was encouraged by the first successful test launches of the R-7 rocket in August and September. These tests proved that the rocket could carry Sputnik 1 into orbit. In late September, Korolev learned that the United States planned to announce a major space achievement during an International Geophysical Year (IGY) conference at the National Academy of Sciences in Washington, D.C., scheduled for October 6, 1957. The Americans were preparing a paper entitled "Satellite Over the Planet," and Korolev suspected that von Braun might attempt to launch a satellite using the Jupiter-C rocket around October 4 or 5, coinciding with this announcement. Determined to beat the Americans, Korolev moved quickly to hasten the launch of Sputnik 1.

The launch vehicle for PS-1 was a modified version of the R-7 rocket, designated 8K71PS number M1-PS. Unlike the earlier R-7 test launches, much of the test equipment and radio gear was stripped from the rocket to streamline the mission. In September 1957, the launch vehicle arrived at the Soviet missile base at Tyura-Tam (later known as Baikonur Cosmodrome) and was prepped for its historic mission at launch site number one.

On Friday, October 4, 1957, at exactly 10:28:34 pm Moscow time, the R-7 rocket carrying Sputnik 1 roared to life. As it ascended from the launch pad, the world's first artificial satellite was on its way into history. The satellite, a small metallic sphere, emitted a steady "beep... beep... beep" from its radio transmitters, which anyone worldwide could pick up with a shortwave radio. Although celebrations at the launch control center were initially subdued, a far-east tracking station at Kamchatka soon confirmed the satellite's successful entry into orbit by detecting its distinctive beeping signals.

Just 95 minutes after the launch, Sputnik 1 completed its first orbit around the Earth, passing over the launch site at Tyura-Tam. The engineers and military personnel picked up its radio signals on the ground, and only then did Korolev and his team celebrate their groundbreaking achievement. Sputnik 1, the world's first artificial satellite, was now orbiting Earth, marking the beginning of the space age and a resounding Soviet victory in the early stages of the Space Race.

This beeping sphere, weighing less than 200 pounds, symbolized far more than a technical triumph—it was a bold statement of Soviet technological superiority. Sputnik 1's success stunned the world and marked a significant turning point in the Cold War, spurring the United States to accelerate its own efforts in space exploration and forever altering the course of human history.

Five days after the launch of Sputnik 1, the world's first artificial satellite, U.S. President Dwight D. Eisenhower addressed the nation, seeking to calm growing concerns about the Soviet Union's achievement. When a reporter asked about potential security threats posed by the satellite, Eisenhower responded confidently: "Now, so far as the satellite itself is concerned, that does not raise my apprehensions, not one iota." He downplayed the significance of Sputnik, asserting that it was a scientific milestone rather than a military threat. According to Eisenhower, the satellite's weight was not indicative of any substantial military capability, and he sought to reassure the public that the Soviet achievement was not a shift in global power.

Despite his public dismissal of Sputnik as a military concern, by 1958 Eisenhower recognized that the United States needed to confront several critical realities. In a candid assessment, he outlined three "stark facts" that America had to face:

1. The Soviet Union had surpassed the United States and the rest of the "free world" in scientific and technological advancements in space.
2. If the Soviet Union maintained its lead in space, it could use this achievement to undermine U.S. prestige and global leadership.
3. Should the Soviets gain superior military capabilities in outer space, it could create an imbalance of power, posing a direct threat to U.S. national security.

Eisenhower urged the nation to meet these challenges with "resourcefulness and vigor." A key aspect of his response focused on the importance of education in bolstering America's scientific and technological future. Reflecting on the role of education in the Soviet Union's recent progress, he stressed that the United States needed to cultivate a generation of scientists. "We need scientists in the ten years ahead," he declared, calling on Americans to scrutinize school curricula and ensure that they met the demands of the rapidly advancing space age.

However, Eisenhower's ability to project confidence was hampered by the secretive nature of his reassurance. Much of his confidence stemmed from clandestine U.S. reconnaissance programs, which provided him with intelligence that was not shared with the public. As a result, he struggled to quell widespread fears that the balance of power had shifted in favor of the Soviet Union. The launch of Sputnik, coupled with concerns over U.S. missile capabilities, caused a temporary dip in Eisenhower's approval ratings, though he eventually recovered politically.

The media played a significant role in amplifying the public's anxiety. In the immediate aftermath of Sputnik's launch, newspapers and television programs sensationalized the event, stirring what became known as a moral panic. The *New York Times* declared the Soviet achievement a major global propaganda victory for communism. American journalist Fred Hechinger reported that the launch of Sputnik led to a surge of media coverage examining the U.S.

education system, further fueling public concern. The constant stream of news reports created a "nation in shock," with journalists frequently exaggerating the dangers posed by the Soviet satellite. Science fiction writer Arthur C. Clarke remarked on October 9, 1957, that with Sputnik's launch, the United States had become "a second-rate power."

Politicians also capitalized on the event, using the fear of Soviet technological superiority to bolster their own ratings. Congress responded to the perceived crisis by dramatically increasing spending on research and development, hoping to close the technological gap. Alarmed by estimates of Soviet rocket strength, both Congress and the executive branch pressured Eisenhower to accelerate missile development, leading to a rapid escalation of the missile race.

While the public's reaction was one of concern, political analyst Samuel Lubell's research found no evidence of widespread panic or hysteria among ordinary Americans. Instead, the panic surrounding Sputnik was largely confined to the political and media elites, who overestimated the immediate threat posed by Soviet advancements. Nonetheless, Sputnik's launch provoked Congress into taking decisive action to improve the U.S. standing in science and technology, marking the beginning of a significant shift in American policy.

Soviet leader Nikita Khrushchev, reflecting on the event, reveled in the perceived fear generated by Sputnik. "Our potential enemies cringe in fright," he boasted, although he acknowledged that the Soviet Union still struggled to achieve the pinpoint accuracy needed for missile strikes. The launch of Sputnik 1 was a profound propaganda victory for the Soviet Union, showcasing its technological prowess and unsettling the West. Though the U.S. would eventually recover from its initial shock, the launch of Sputnik fundamentally altered the course of the Space Race, intensifying the rivalry between the superpowers.

In the week following the launch of Sputnik in October 1957, the United States was jolted by a wave of anxiety. The surprise Soviet achievement triggered an avalanche of government funding for scientific research and education. John Jefferies, an astronomer at the High Altitude Observatory, recalled the sudden influx: "The week after Sputnik went up, we were digging ourselves out of this avalanche of money that suddenly descended." The launch set off a series of sweeping American initiatives aimed at reclaiming leadership in space, ranging from defense and education to technological innovation.

At the forefront of these initiatives was the U.S. Navy's Project Vanguard, which was tasked with launching an American satellite. However, the first success came from the Army's Explorer program. On January 31, 1958, the United States launched its first satellite, Explorer 1, marking America's official entry into the space race. Shortly after, in February 1958, President Dwight D. Eisenhower authorized the creation of the Advanced Research Projects Agency (ARPA), later known as the Defense Advanced Research Projects Agency (DARPA), to spearhead the development of emerging technologies for national defense.

The Sputnik crisis also reshaped American education. The fear of falling behind technologically spurred an unprecedented focus on STEM (Science, Technology, Engineering, and Math) fields, as well as foreign languages and cultural studies, all seen as essential to national security. This focus on education became a central component of America's Cold War strategy, driven by the belief that technological supremacy would be key to global dominance.

The launch of Sputnik 2, which carried a payload of more than 500 kg, shocked the world and cemented the Soviet Union's lead in rocket technology. The CIA, caught off guard, initially estimated the launch weight at a staggering 500 metric tons, requiring an initial thrust of over 1,000 tons. The agency speculated that the Soviets had used a three-stage rocket to achieve this feat. In a secret report, the CIA described the launch of two earth satellites as a "stupendous scientific achievement" and concluded that the USSR had likely perfected an intercontinental ballistic missile (ICBM) capable of striking any target with precision.

However, this assessment was based on a significant miscalculation. In reality, the Soviet rocket had a launch weight of 267 metric tons with an initial thrust of 410 tons and operated with one and a half stages, not the three stages the CIA had assumed. The misjudgment stemmed from the CIA's reliance on U.S. Atlas rocket parameters, which at the time had a launch weight of 82 tons, an initial thrust of 135 tons, and could only carry a maximum payload of 70 kg into low Earth orbit.

Part of the Soviet rocket's success was attributed to innovative concepts introduced by German rocket scientists, including Helmut Gröttrup, who had been working on Gorodomlya Island after World War II. These innovations included rigorous weight-saving measures, efficient control of residual fuel quantities, and a reduced thrust-to-weight ratio of 1.4, compared to the typical factor of 2 used in Western rockets. The CIA had been aware of such developments as early as January 1954, when they interrogated Gröttrup following his return from the USSR. However, at the time, his insights were largely dismissed.

The CIA's misjudgment highlighted the agency's underestimation of Soviet technological capabilities. The successful launch of Sputnik 2 not only demonstrated the USSR's lead in space exploration but also signaled that the Soviet Union had made significant advances in missile technology, escalating Cold War tensions and fueling the American drive to catch up in the space race.

Even decades later, the echoes of Sputnik lingered. The sense of urgency it ignited continued to shape American attitudes toward education, innovation, and national security. The rise of Japan as a technological competitor in the 1980s, for instance, revived fears of a "technology gap" reminiscent of the Sputnik era, as American leaders sought to rekindle the spirit of innovation that had driven the country's response to the Soviet challenge.

In the end, the launch of Sputnik not only marked the beginning of the space race but also catalyzed a period of intense scientific, educational, and military advancements in the United States. It set the stage for the United States' eventual triumph in the race to the Moon and cemented space exploration as a defining arena of Cold War competition.

In Britain, the launch of Sputnik in 1957 was met with a mixture of surprise and excitement as the nation realized it was witnessing the dawn of the Space Age. The event was a striking reminder of Britain's diminished global influence, particularly in contrast to the superpowers of the United States and the Soviet Union, whose technological rivalry now dominated the international stage. This moment also underscored Britain's role in the broader Cold War narrative, as its geopolitical significance shifted in the face of the rapidly advancing space race.

British media captured the public sentiment well. The *Daily Express* confidently predicted that the United States would respond swiftly to regain technological superiority, stating, "The result will be a new U.S. drive to catch up and pass the Russians in the sphere of space exploration. Never doubt for a moment that America will be successful." This reflected a widespread belief that the U.S., with its industrial might and resources, would inevitably outpace the Soviet Union in space exploration.

The Sputnik crisis also had significant diplomatic and defense ramifications for Britain. It contributed to the U.S.–UK Mutual Defence Agreement of 1958, which solidified military cooperation between the two nations, particularly in nuclear technology and defense systems. This agreement further aligned Britain with the United States in the strategic competition against the Soviet Union, emphasizing the interconnected nature of the Cold War's military, technological, and geopolitical struggles.

The launch of Sputnik may have signaled a shift in world power dynamics. Still, it also reaffirmed Britain's close alliance with the United States in the emerging space race, tying its future more tightly to the technological and military advancements that would come to define the latter half of the 20th century.

The Soviet Union's success with Sputnik caused deep concern across the United States. Bernard Baruch, a prominent American economist, expressed the nation's anxiety in an open letter titled "The Lessons of Defeat," published in the *New York Herald Tribune*. Baruch wrote, "While we devote our industrial and technological power to producing new model automobiles and more gadgets, the Soviet Union is conquering space... It is Russia, not the United States, who has had the imagination to hitch its wagon to the stars and the skill to reach for the moon and all but grasp it. America is worried. It should be." His words reflected the growing fear that the Soviet Union was not only ahead in the space race but could also wield space technology as a strategic military advantage.

In response to the Soviet Union's successful launch of Sputnik 1 in October 1957, President Dwight D. Eisenhower accelerated U.S. efforts in space, pushing forward the timeline for Project Vanguard, a program intended to launch America's first satellite. However, this haste led to a highly publicized failure on December 6, 1957, when the Vanguard rocket exploded just seconds after liftoff at Cape Canaveral, Florida. The event was an embarrassment for the United States, and the press quickly seized the opportunity to mock the failure, coining derisive nicknames like "Flopnik," "Kaputnik," "Dudnik," and "Stayputnik." Internationally, the failure was even used as political ammunition, with a Soviet delegate to the United Nations sarcastically offering American assistance through their "technical aid program for backward nations."

Amid this public humiliation, Wernher von Braun's team, working on the Redstone missile under the U.S. Army, was finally given the green light to proceed with their Jupiter-C rocket. On January 31, 1958, almost four months after Sputnik's debut, the United States successfully launched its first satellite, Explorer 1, aboard the Juno I rocket, derived from the Redstone missile. Explorer 1, weighing only 30.66 pounds (13.91 kg), carried instruments designed by Dr. James Van Allen, a prominent physicist at the University of Iowa, including a micrometeorite gauge and a Geiger-Müller tube. These instruments made a groundbreaking discovery—detecting the Van Allen radiation belts, doughnut-shaped zones of high-energy radiation encircling Earth. Despite its small size, Explorer 1 was a triumph for American space science, providing critical data about the space environment and marking the nation's official entry into the space race.

Recognizing that a more comprehensive and coordinated effort was necessary to compete with the Soviet Union, President Eisenhower took decisive action. On April 2, 1958, he proposed to Congress the creation of a civilian space agency to spearhead nonmilitary space exploration efforts. The response was swift. Led by Senate Majority Leader Lyndon B. Johnson, Congress passed the National Aeronautics and Space Act, which Eisenhower signed into law on July 29, 1958. This landmark legislation transformed the National Advisory Committee for Aeronautics (NACA) into the National Aeronautics and Space Administration (NASA), entrusting it with the responsibility for all civilian space activities. To ensure strategic coordination, a Civilian-Military Liaison Committee was also established to align civilian and military space programs, securing the nation's interests in the new frontier.

The establishment of NASA marked a turning point for the American space program, bringing a sense of unity and purpose to U.S. space efforts. In a further strategic move, on October 21, 1959, Eisenhower approved the transfer of the U.S. Army's remaining space-related projects, including von Braun's team, to NASA. This transfer formalized the creation of NASA's George C. Marshall Space Flight Center in Huntsville, Alabama, on July 1, 1960, with von Braun appointed as its first director. One of NASA's immediate objectives was the development of the powerful Saturn rocket family, which would later play a critical role in enabling the United States to achieve ambitious goals in space exploration.

By the early 1960s, the U.S. had not only closed the technological gap with the Soviet Union but had also laid the groundwork for the remarkable achievements that followed, including the Apollo program and the eventual Moon landing in 1969. The creation of NASA, coupled with the transfer of military expertise into civilian hands, marked the United States' emergence as a spacefaring nation, driven by Cold War rivalry that had sparked the space race.

One of the most significant developments in this period came with the signing of the National Aeronautics and Space Act, but it was not the only critical legislative move. Less than a year after Sputnik's launch, the U.S. Congress passed the National Defense Education Act (NDEA) in 1958, providing billions of dollars in federal funding to bolster education, particularly in science, technology, engineering, and mathematics (STEM) fields. By 1960, federal spending on education had increased sixfold, reflecting the urgent need to cultivate a new generation of scientists and engineers to lead America's efforts in the emerging space age. These initiatives not only fueled technological advancements but also ensured the long-term sustainability of the U.S. space program, paving the way for the historic achievements that lay ahead.

As the space race intensified, the U.S. found itself in a geopolitical and technological battle with the Soviet Union. When campaigning for the presidency in 1960, John F. Kennedy criticized the "missile gap" between the U.S. and the Soviet Union and promised to expand America's missile arsenal. Though initially skeptical of manned space missions, Kennedy was forced to reassess after the Soviet Union sent the first human, Yuri Gagarin, into space on April 12, 1961. Gagarin's historic flight stunned the world and spurred Kennedy to set an ambitious goal—landing a man on the Moon before the decade's end. He declared, "If the Soviets control space, they can control the Earth."

The launch of Sputnik had far-reaching consequences beyond space exploration. It prompted a profound transformation in U.S. science policy and funding. NASA soon became one of the largest sources of federal support for academic research. By the mid-1960s, nearly 10% of federal funding for university research came from NASA, catalyzing the growth of American scientific innovation.

Chapter 2 – The Space Race Begins

The impact extended to defense as well. The U.S. rapidly expanded its missile defense and space weapons programs with proposals for anti-ballistic missile systems and investments in missile technology. Simultaneously, the National Science Foundation (NSF) saw its budget increase dramatically, from $34 million in 1958 to nearly $500 million by 1968, fueling advancements in science and technology.

On July 29, 1955, the United States formally announced its intention to launch an artificial satellite during the International Geophysical Year, which spanned from July 1, 1957, to December 31, 1958. This announcement marked the beginning of the U.S. effort to enter space exploration, as the country sought to demonstrate its technological prowess on the global stage. The International Geophysical Year, a scientific event aimed at encouraging international cooperation in the study of Earth sciences, provided the perfect platform for showcasing advancements in space technology. The United States planned to contribute to this effort by launching a satellite that would orbit Earth and collect valuable scientific data.

Just over a month later, on August 30, 1955, the Soviet Union responded with its own bold step. The Soviet space commission approved a plan to launch a significantly larger satellite using their newly developed R-7 intercontinental ballistic missile (ICBM), which had the capability of lifting a one-ton payload into orbit. This decision, approved under the leadership of Sergei Korolev, the chief designer of Soviet space programs, set the stage for what would become a fierce technological competition between the two superpowers. The R-7 ICBM, initially developed for military purposes, would soon be repurposed as the launch vehicle for *Sputnik 1*, the world's first artificial satellite.

These parallel announcements from the United States and the Soviet Union marked the formal beginning of the Space Race. Both nations, driven by Cold War tensions and the desire to demonstrate their technological superiority, were now locked in a race to launch the first satellite into space.

In 1957, the Soviet Union rapidly accelerated its dominance in both military technology and space exploration, marking a series of historic milestones that would change the course of the Space Race.

R-7 Semyorka

First Intercontinental Ballistic Missile (Icbm); Fully Operational September 1957

The R-7 Semyorka, officially designated the GRAU index 8K71, was a groundbreaking Soviet missile developed during the Cold War. It holds the distinction of being the world's first intercontinental ballistic missile (ICBM). Developed between 1953 and 1957, the R-7 made 28 launches between 1957 and 1961. A variant known as the R-7A was operational from 1960 until 1968. Unveiled to the West following its first successful launch, the R-7 was assigned the NATO reporting name SS-6 Sapwood. Beyond its military origins, the R-7 became the foundation for the Soviet Union's space program, launching *Sputnik 1*, the first artificial satellite, and forming the basis for the R-7 family of space launchers, which included the *Sputnik, Luna, Molniya, Vostok, Voskhod*, and later *Soyuz* rockets. The R-7's legacy continues today, as modified versions of this missile have become the world's most reliable space launch vehicles.

The R-7 Semyorka was an imposing machine. It measured 34 meters in length, with a diameter of 10.3 meters, and had a launch mass of 280 metric tons. The missile was designed to carry a single thermonuclear warhead with a yield of 3 megatons of TNT over a range of up to 8,000 kilometers. Its accuracy, with a circular error probable (CEP) of around 5 kilometers, was impressive for its time.

The R-7 used a unique configuration, consisting of a central core stage powered by a single engine surrounded by four strap-on boosters. Each booster had two vernier thrusters to assist with steering, while the central core engine provided main propulsion. The missile's innovative design eliminated the need for jet vanes, instead relying on vernier thrusters for precise control. Propellant tanks for the boosters and the central core were separate, and the development team had to design a sophisticated system to synchronize the propellant consumption across the entire missile.

One of the R-7's notable features was its launch system. Rather than launching from a horizontal platform, the missile was suspended by trusses, which bore both its vertical weight and withstood horizontal wind forces. This setup reduced the risk of the missile collapsing under its own weight or being toppled by strong gusts of wind.

The R-7's development began in 1953, and it was led by Soviet rocket designer Sergei Korolev and his team at OKB-1 in Kaliningrad (now Korolyov). The initial design was based on earlier Soviet research on multi-stage rockets, particularly a "rocket packet" design proposed by Mikhail Tikhonravov in 1947. Korolev's team further refined the concept, selecting a configuration that featured a central core stage with four strap-on boosters, which proved to be the optimal design for achieving the desired range and payload capacity.

The engine development, led by Valentin Glushko's OKB-456, utilized a new approach of four combustion chambers with a single turbopump, significantly increasing thrust and reducing overall engine weight. This configuration powered both the central core and the strap-on boosters.

By 1957, the design had evolved into the R-7 Semyorka, capable of carrying a 5.5-ton nuclear warhead. The USSR Council of Ministers officially approved its development on May 20, 1954.

To accommodate the size and complexity of the R-7, the Soviets needed a new testing site. Kapustin Yar, their existing launch facility, was insufficient. As a result, construction began on what would become the Baikonur Cosmodrome in Kazakhstan in 1955.

The first flight-ready R-7 was delivered to Baikonur in May 1957, and the missile's inaugural launch occurred on May 15. However, this first attempt failed when a fire broke out in one of the strap-on boosters, causing the missile to disintegrate 88 seconds after liftoff. The second attempt on June 11 also failed due to an electrical short, causing the missile to roll uncontrollably before breaking apart 33 seconds into the flight.

Success came on August 21, 1957, when the R-7 completed a 6,000-kilometer flight, reaching its target near Kamchatka. However, the dummy warhead disintegrated in the atmosphere. Another successful test on September 7 confirmed the missile's long-range capabilities, though warhead re-entry issues persisted.

On August 26, 1957, the Soviet news agency TASS announced to the world that the USSR had successfully tested the world's first ICBM. The R-7 had successfully achieved what no other country had done at that time—develop a missile capable of delivering a nuclear warhead across continents.

Though the R-7 was an effective demonstration of Soviet missile technology, it proved impractical as a strategic weapon. The R-7 required almost 20 hours of preparation before launch and could not be left on alert for extended periods due to its cryogenic fuel. Moreover, the missile's large launch complexes were easily detectable by U.S. spy planes, leaving them vulnerable to a preemptive strike in the event of a conflict.

Despite these limitations, the R-7 found new life as the foundation for the Soviet space program. On October 4, 1957, a modified R-7 launched *Sputnik 1*, the world's first artificial satellite, into orbit, marking the beginning of the Space Race. A month later, *Sputnik 2* carried Laika, the first animal in space, aboard an R-7 variant. These early successes in space exploration solidified the R-7's role as a launch vehicle, and by the end of 1959, the R-7 had successfully placed several satellites into orbit.

The R-7A, an improved version of the original missile, became operational in 1960, with an extended range of 12,000 kilometers. However, by 1968, the R-7 had been phased out of military service in favor of more practical second-generation missiles like the R-16.

Although the R-7 was retired from military service, its legacy as the basis for the Soviet and later Russian space launch vehicles remains strong. The R-7's design became the foundation for the *Vostok, Voskhod, Soyuz,* and *Molniya* families of launchers. As of 2024, its descendants, including the Soyuz-U and Soyuz-FG, remain in service, making the R-7 one of the most reliable and enduring space launch systems in history. Having launched over 1,840 times, the R-7 stands as a testament to Soviet engineering and its critical role in humanity's journey into space.

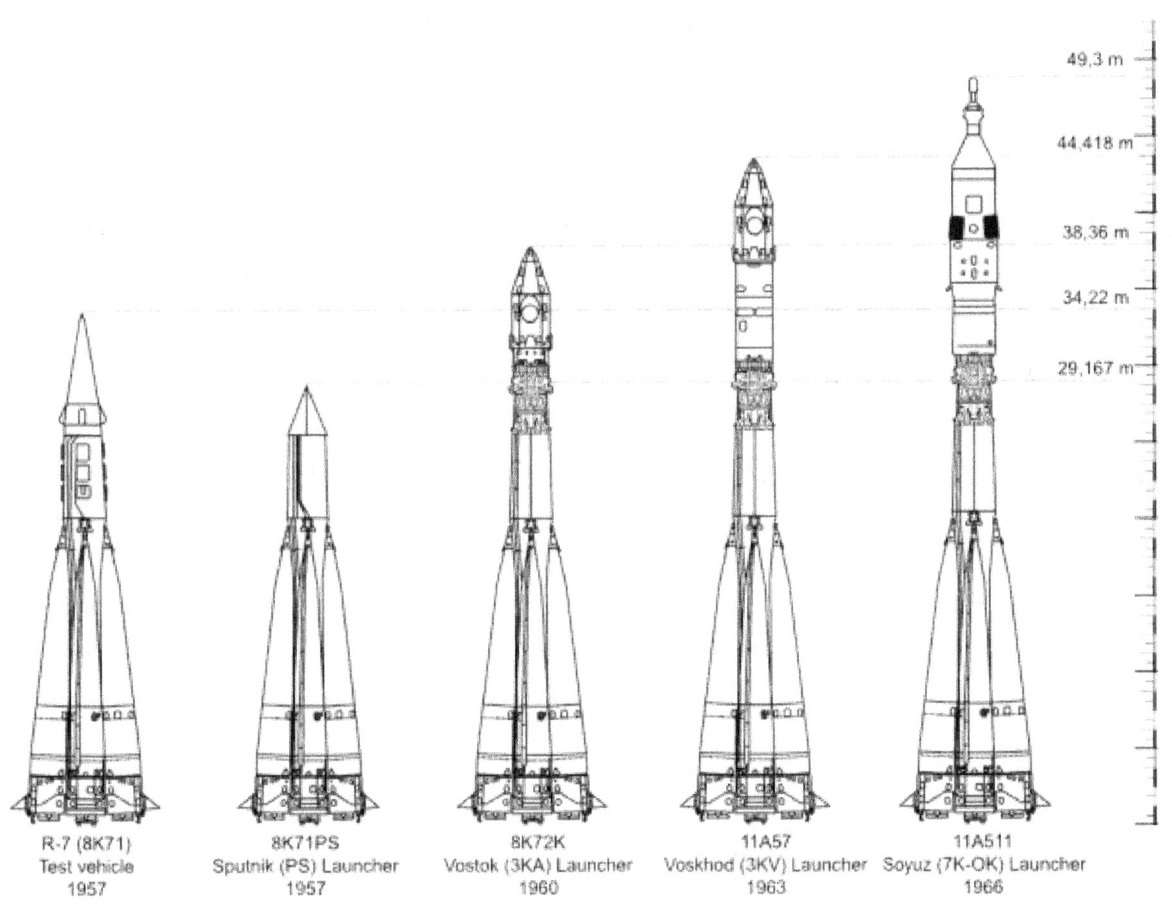

Sputnik 1

First artificial satellite
First man-made signals from orbit

Just a few months later, on October 4, 1957, the Soviet Union launched *Sputnik 1*, the first artificial satellite to orbit the Earth. On October 4, 1957, the Soviet Union shocked the world by launching *Sputnik 1*, the first artificial Earth satellite, marking a significant milestone in the space race. Officially designated as *Спутник-1* (Russian for "Satellite-1"), the small, polished sphere was sent into an elliptical low Earth orbit from the Baikonur Cosmodrome in Kazakhstan, then part of the Soviet Union. This launch was part of the International Geophysical Year, a global initiative for scientific collaboration.

The Soviet R-7 rocket, initially designed as an intercontinental ballistic missile (ICBM), was adapted to place *Sputnik 1* into orbit. The decision to develop this powerful rocket had been made in 1954, as part of Soviet plans to ensure their technological superiority during the Cold War. The R-7, with excess thrust to carry heavy payloads such as hydrogen bombs, was also capable of launching objects into space. Known by NATO as the T-3 or Type A, the R-7 was a revolutionary rocket at the time.

At exactly 19:28 UTC, the rocket lifted off from Site No.1, carrying *Sputnik 1* into orbit. Despite some technical difficulties, including a premature fuel depletion, the launch succeeded, with the core stage and satellite reaching an initial orbit of 223 kilometers by 950 kilometers, just shy of the intended 1,450 kilometers. The satellite began transmitting a steady "beep-beep" signal on 20.005 and 40.002 MHz, signals that were picked up by radio operators worldwide. These signals continued for 22 days, until the satellite's batteries were depleted on October 26, 1957.

Sputnik 1 itself was a small, highly polished sphere, measuring only 58 centimeters in diameter and weighing 83.6 kilograms. It orbited the Earth every 96 minutes at a velocity of approximately 8 kilometers per second (about 18,000 miles per hour), completing 1,440 orbits during its three-month stay in space. The satellite burned up in Earth's atmosphere on January 4, 1958, after traveling approximately 70 million kilometers.

The Soviet Union's achievement in launching *Sputnik 1* had far-reaching political, scientific, and military implications. It signaled the USSR's technological prowess and opened the door to space exploration. The satellite's launch vehicle, the R-7 rocket, was a testament to Soviet engineering, having been designed and tested under the direction of Sergei Korolev, the chief architect of the Soviet space program.

Korolev and his team watched the launch from the range with great anticipation. About 90 minutes after liftoff, confirmation came that *Sputnik 1* had successfully completed its first orbit. This monumental moment led to Soviet Premier Nikita Khrushchev being informed of the success. Shortly thereafter, the Soviet Telegraph Agency (TASS) announced the achievement to the world, marking the start of the Space Age.

For the Soviet Union, the launch of *Sputnik 1* was not only a scientific breakthrough but also a potent symbol of ideological victory over the West. Soviet citizens were encouraged to listen to the satellite's signals on their radios, and though *Sputnik 1* was difficult to see with the naked eye, the 26-meter core stage of the R-7 rocket, which followed in orbit behind the satellite, was easily visible.

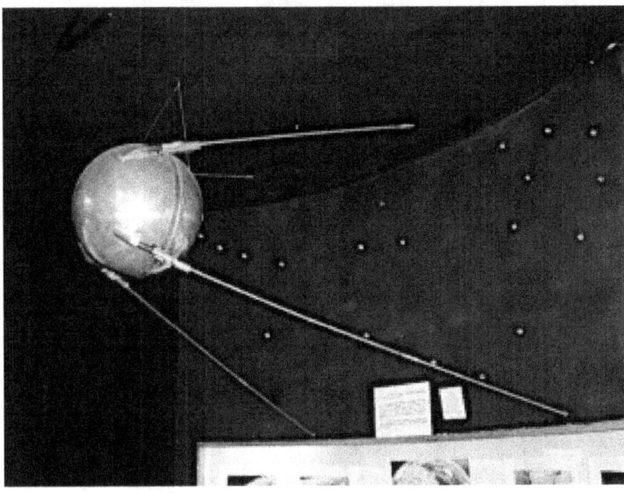

Sputnik 1

While *Sputnik 1* orbited silently in space, its success stirred immediate reactions across the globe. In the United States, it led to the rapid acceleration of American space efforts, culminating in the establishment of NASA in 1958. The Soviet Union, meanwhile, pressed forward with further space missions, spurred by Khrushchev's demand for another spectacular achievement. This led to the launch of *Sputnik 2* in November 1957, carrying the first living being, the dog Laika, into orbit.

Sputnik 2

First mammal (the dog Laika) in orbit around Earth

Building on this success, the Soviet Union followed with another groundbreaking achievement just a month later. On November 3, 1957, the Soviet Union achieved another monumental milestone in the space race by launching *Sputnik 2*, the second artificial satellite to orbit the Earth. Unlike its predecessor, *Sputnik 2* carried a living passenger: Laika, a small, mixed-breed dog who became the first animal to orbit the planet. This mission marked a significant leap forward in both technological prowess and scientific discovery, even though the outcome for Laika was tragic.

Sputnik 2

Launched just a month after *Sputnik 1*, *Sputnik 2* was a more complex spacecraft, weighing approximately 500 kilograms and standing four meters high. Its cone-shaped design included multiple compartments to house a telemetry system, scientific instruments, and a temperature-regulated cabin for Laika. The total mass in orbit, including the attached rocket core, was 7.79 tons, a dramatic increase compared to *Sputnik 1*.

The spacecraft was launched atop a modified R-7 intercontinental ballistic missile (ICBM) from the Baikonur Cosmodrome, using a similar configuration to that which had successfully launched *Sputnik 1*. However, unlike the first satellite, *Sputnik 2* was designed to carry scientific instruments capable of continuously observing the Earth's atmosphere and solar radiation, in addition to its biological payload. This mission aimed to gather data on space's effects on living organisms and contributed to the understanding of cosmic rays and solar activity.

Laika, whose name translates to "Barker," was chosen for the mission because of her calm temperament. Weighing approximately six kilograms, she had been selected from a group of dogs already trained for suborbital flights on Soviet sounding rockets. Laika, a stray mongrel from the streets of Moscow, was housed in a small, padded cabin that provided just enough room for her to sit or lie down.

The cabin was fitted with life-support systems that provided oxygen, food in a gelatinized form, and monitoring devices for her vital signs. Wires were surgically attached to her body to track her heartbeat, respiration, and blood pressure throughout the mission. The experiment, which monitored Laika's vital signs, aimed to prove that a living organism could survive being launched into orbit and continue to function under weakened gravity and increased radiation, providing scientists with some of the first data on the biological effects of spaceflight.

The launch of *Sputnik 2* was timed to coincide with the 40th anniversary of the Bolshevik Revolution, a date emphasized by Soviet Premier Nikita Khrushchev. Soviet engineers had just three weeks to prepare the spacecraft following the successful launch of *Sputnik 1*, forcing them to repurpose parts from earlier missions, including payload containers used in suborbital tests with animals. The mission was executed under tight deadlines, and no provision was made for Laika's recovery. Soviet officials initially claimed Laika would be euthanized with poisoned food, but it was later revealed that she died from overheating a few hours into the mission, likely due to a failure in the temperature control system.

Despite Laika's untimely death, *Sputnik 2* continued to transmit scientific data for several days. The spacecraft's cosmic ray detector, developed by the Lebedev Institute of Physics, reported a significant increase in charged particles at higher latitudes, suggesting that *Sputnik 2* was beginning to detect the lower edges of what would later be known as the Van Allen radiation belts. However, the satellite's X-ray and ultraviolet instruments failed to return usable data due to oversaturation from cosmic radiation.

Laika

The mission concluded on April 14, 1958, when *Sputnik 2* reentered Earth's atmosphere and burned up after completing 2,370 orbits over 162 days. Its reentry was widely observed, with reports of the glowing debris stretching from New York to the Amazon.

In the geopolitical context, the launch of *Sputnik 2* had a profound impact. Compared to the earlier Sputnik 1, the satellite's significantly larger mass emphasized the Soviet Union's lead in space technology and missile capability, which alarmed the United States. President Eisenhower's administration, while remaining outwardly calm, faced growing pressure to accelerate American space efforts. The launch also spurred debate worldwide about the ethics of animal testing, with protests from animal rights groups, especially in the United Kingdom and the United States, over Laika's treatment.

While *Sputnik 2* did not lead directly to major scientific breakthroughs, it laid the groundwork for future missions and demonstrated the feasibility of sending living beings into space, paving the way for human spaceflight. The mission also underscored the competitive nature of the space race, as both superpowers sought to claim dominance in the final frontier.

The US Vanguard 1

First solar-powered satellite

On March 17, 1958, the United States launched Vanguard 1, the fourth artificial Earth-orbiting satellite, following the Soviet Union's *Sputnik 1*, *Sputnik 2*, and the U.S.'s *Explorer 1*. This 1.46 kg aluminum sphere, measuring only 6 inches in diameter, marked a significant milestone in space exploration as the first satellite to be powered by solar energy. Its mission was part of Project Vanguard, aimed at testing a three-stage vehicle's launch capabilities and studying the space environment's effects on satellites in Earth orbit.

Vanguard 1 was equipped with two transmitters: a 10 mW mercury-battery-powered transmitter operating at 108 MHz and a 5 mW transmitter at 108.03 MHz, powered by six solar cells mounted on its surface. The satellite's unique design, with six spring-actuated antennas extending from its body, enabled it to transmit signals for tracking purposes and measure the total electron content between the satellite and ground stations. These signals played a crucial role in geodetic measurements, helping scientists understand the shape and size of the Earth with greater precision.

The satellite was launched into an elliptical orbit with an apogee of 3,969 kilometers and a perigee of 654 kilometers, with an inclination of 34.25 degrees. Originally, it was expected that Vanguard 1 would remain in orbit for 2,000 years. However, due to solar radiation pressure and atmospheric drag during periods of high solar activity, its expected orbital lifetime was reduced to approximately 240 years. The satellite continues to orbit Earth today, making it the oldest human-made object still in space.

Vanguard 1 transmitted data for over six years, providing vital information about atmospheric density. Its symmetrical shape made it particularly useful for determining the density of the upper atmosphere by tracking the slight deviations in its orbit due to atmospheric drag. These observations allowed scientists to measure atmospheric pressure and other parameters at various altitudes and solar conditions, significantly contributing to our understanding of the Earth's atmosphere.

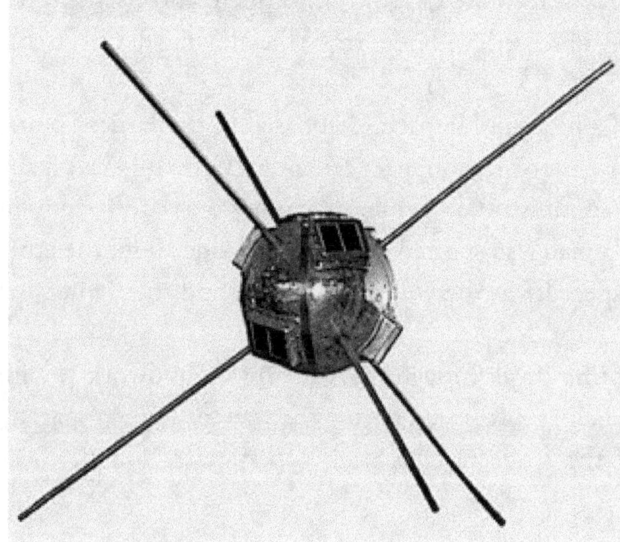

Vanguard 1 Satellite

Although overshadowed by earlier Soviet achievements, the launch of Vanguard 1 was a technical triumph for the United States and its space program. The satellite's small size and innovative use of solar power earned it the nickname "the grapefruit satellite," as Soviet Premier Nikita Khrushchev mockingly referred to it. Despite its diminutive size, Vanguard 1 provided critical data and remained a symbol of early American efforts in space exploration.

Though communications with the satellite were lost in 1964, its presence in orbit endures as a relic of the early days of the Space Race. Alongside the upper stage of its launch vehicle, Vanguard 1 will continue to orbit Earth for centuries, potentially burning up in the atmosphere in the late 22nd century. As the world moves further into the age of space exploration, some have suggested that Vanguard 1 might one day be retrieved as a historical artifact, embodying humanity's first steps into the cosmos.

Luna 1

First lunar spacecraft (fly-by)
First rocket engine restart in Earth orbit
First spacecraft to leave Earth's orbit
First spacecraft on an escape trajectory from Earth

First spacecraft in heliocentric orbit

Luna 1, also known as Mechta (Russian for "Dream"), holds a significant place in space exploration as the first spacecraft to escape Earth's orbit and approach the Moon. Launched on January 2, 1959, as part of the Soviet Luna program, Luna 1 marked a significant milestone in the Space Race between the United States and the Soviet Union. It was the first human-made object to enter heliocentric orbit, a remarkable achievement that laid the groundwork for future interplanetary missions.

The spacecraft's mission was initially intended to impact the Moon, carrying metallic pennants emblazoned with the Soviet coat of arms. However, a malfunction in the ground-based control system caused an error in the burn time of the upper stage rocket, resulting in Luna 1 missing its target by 5,900 kilometers (3,700 miles). Despite this, Luna 1 became the first human-made object to escape Earth's gravitational influence and was aptly dubbed "Artificial Planet 1" by Soviet scientists. It was later renamed Mechta, symbolizing humanity's aspirations to explore the cosmos. The spacecraft was also called the "First Cosmic Ship" due to its achievement of Earth escape velocity, further cementing its place in space history.

Luna 1 was a spherical satellite powered by mercury-oxide batteries and silver-zinc accumulators. It measured 1.22 meters (4 feet) in diameter and weighed 361.3 kilograms (797 pounds) at launch. Its exterior featured five antennas designed for communication and a scientific payload intended to gather data on the Moon and the surrounding space environment. The spacecraft lacked a propulsion system, relying solely on its launch vehicle for movement.

Among its instruments were a flux-gate magnetometer capable of measuring magnetic fields, two micrometeorite detectors developed by Soviet scientist Tatiana Nazarova, and ion traps created by Konstantin Gringauz to measure solar wind and plasma. The scientific payload also included gas-discharge Geiger counters, a sodium-iodide scintillation counter, and a Cherenkov detector, all designed to study cosmic rays and the surrounding space environment. The launch vehicle's upper stage carried a scintillation counter and one kilogram of sodium, which would later be released in space as part of a scientific experiment.

Luna 1 was launched from the Baikonur Cosmodrome on January 2, 1959, atop a Luna 8K72 rocket. The first three stages of the rocket operated nominally, but a manual error in controlling the engine burn of the Block E upper stage caused the spacecraft to miss its intended lunar impact. Instead of crashing into the Moon, Luna 1 passed within 5,995 kilometers (3,725 miles) of the lunar surface on January 4, 1959, and continued on its trajectory into heliocentric orbit.

The spacecraft ran out of battery power on January 5, 1959, when it was approximately 597,000 kilometers (371,000 miles) from Earth. Despite this, the mission achieved several scientific firsts, including detecting solar wind and observing the Van Allen radiation belts. Luna 1's escape from Earth's orbit and entry into heliocentric orbit were groundbreaking achievements in the context of the Cold War space race, signaling the Soviet Union's growing technological capabilities in space exploration.

Luna 1

One of the most striking experiments conducted by Luna 1 involved the release of a one-kilogram cloud of sodium gas on January 3, 1959, at 113,000 kilometers (70,000 miles) from Earth. This created an artificial comet, visible from Earth as a glowing orange trail, which was used to track the spacecraft's trajectory and study the behavior of gases in space. The sodium cloud experiment was one of the first demonstrations of how humans could observe space phenomena from Earth, adding a new dimension to space exploration.

Luna 1 also significantly contributed to studying Earth's outer radiation belts. As it passed through the Van Allen belts, its instruments detected the presence of high-energy particles in the outer belt, providing new insights into the structure and intensity of the radiation surrounding Earth. Additionally, Luna 1 conducted the first direct observations of solar wind, a stream of charged particles emitted by the Sun. The ionized plasma concentration was measured at altitudes between 20,000 and 150,000 kilometers from Earth, providing the first quantitative data on this phenomenon.

Luna 1's mission was met with skepticism in some parts of the world. Western scientists, unable to receive the spacecraft's transmissions until it had traveled beyond 170,000 kilometers from Earth, questioned the veracity of Soviet claims. However, subsequent hearings in the United States House of Representatives, held in May 1959, affirmed the Soviet mission's technical sophistication. The mission's success, particularly its use of advanced guidance technology, was acknowledged as a significant achievement.

Discoverer 1

First satellite in a polar orbit

The Discoverer program was a significant element in the early days of space reconnaissance, marking the beginning of the U.S. effort to gather critical intelligence during the Cold War. As part of the top-secret CORONA program, Discoverer 1 was the first in a series of satellites designed to prototype a sophisticated film-return photographic surveillance satellite system. Launched on February 28, 1959, aboard a Thor-Agena A rocket from Vandenberg Air Force Base in California, Discoverer 1 represented the cutting edge of space technology. This satellite, however, was not equipped with a camera or film capsule, as its primary mission was to test the satellite's propulsion, guidance systems, and the feasibility of achieving polar orbit, a crucial orbit for reconnaissance purposes.

At the time of its launch, Discoverer 1 was notable for being the first satellite to be launched toward the South Pole in an attempt to achieve polar orbit. A polar orbit would enable the satellite to pass over every point on Earth, providing a unique vantage point for reconnaissance purposes. Although Discoverer 1 was believed to have achieved orbit, the mission was plagued by communication difficulties, with signals being lost shortly after launch. A declassified CIA report suggested that "most people believe the Discoverer 1 landed somewhere near the South Pole," leaving its exact fate a mystery. Despite these challenges, the mission laid the groundwork for future successes.

Discoverer 1 Satellite

The Advanced Research Projects Agency (ARPA) of the U.S. Department of Defense managed the Discoverer program in collaboration with the U.S. Air Force. Its primary objective was to develop a reconnaissance satellite capable of capturing photographic evidence of Soviet military activities, particularly the production and deployment of long-range bombers and ballistic missiles. At the height of the Cold War, this intelligence was critical for understanding Soviet military capabilities. The program aimed to replace the Lockheed U-2 spy plane, vulnerable to Soviet anti-aircraft defenses, with a space-based surveillance system that could operate with relative impunity.

In addition to its military applications, the Discoverer program had significant secondary uses. It contributed to the U.S. government's mapping and charting efforts and helped develop satellite subsystems critical for human space exploration. Officially, the program was presented to the public as a project to test large satellites for communication purposes and to explore the possibility of placing humans in space. In reality, the Discoverer series was a crucial step in developing America's reconnaissance capabilities.

Discoverer 1 was a 5.73-meter long, cylindrical satellite with a diameter of 1.52 meters, capped by a conical nosecone. The satellite was built using a magnesium casing and contained a modest payload of 18 kilograms, primarily communication and telemetry equipment. It carried a high-frequency, low-power beacon transmitter for tracking and a radar beacon transmitter to receive commands and enable long-range radar tracking. The satellite was equipped with 15 telemetry channels to monitor roughly 100 aspects of spacecraft performance, though unlike later Discoverer missions, it lacked a camera or film-return system.

Discoverer launch

The satellite was launched atop a Thor-Agena rocket, a pioneering combination that would become the first two-stage rocket capable of sending a payload into orbit. The Thor first stage, powered by liquid oxygen and RP-1 propellants, accelerated the spacecraft to an altitude where the second-stage Agena ignited, using hypergolic fuels to propel Discoverer 1 into a polar orbit. The Thor-Agena combination proved to be a reliable launch system, forming the backbone of early U.S. satellite missions.

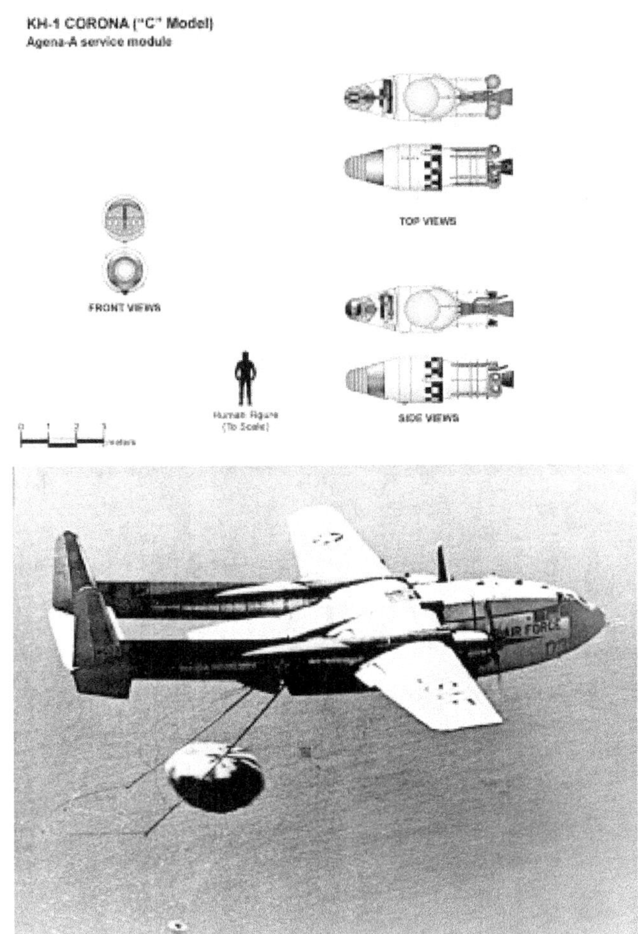

Fairchild_C-
119J_Flying_Boxcar_recovers_CORONA_Capsule_196

The launch crew for Discoverer 1 consisted of five young men who played crucial roles in the success of this historic mission. The crew included James Boyle, Mark Jonah, John Lane, Jack Shields (the Test Conductor), and Erwin Warshawsky. Jack Shields, at 30 years old, was the eldest among them, while the others were in their early to mid-twenties. These men had previously made history in 1958 by launching the first Intermediate-Range Ballistic Missile (IRBM) from the West Coast of the United States, marking a significant achievement in the U.S. missile program.

In the years following the Discoverer 1 mission, Mark Jonah, one of the key members of the launch crew, rose to prominence within the U.S. weapons program. By 1961, at just 27 years old, Jonah became the youngest Test Conductor for the Titan II Intercontinental Ballistic Missile (ICBM) system at Vandenberg Air Force Base. His responsibilities included overseeing the integration of all systems involved in launching a fully operational ballistic missile, from flight control to guidance and propulsion systems. Jonah's leadership in these early days of the U.S. space and missile programs exemplified the rapid advancements and the high-stakes nature of the Cold War-era space race.

The Discoverer series, initiated in 1959, significantly advanced early space reconnaissance technology as part of the Space Race between the United States and the Soviet Union. These missions served both as research and development (R&D) projects and as operational flights under the cover name "Discoverer," eventually leading to the highly classified Corona satellite program.

The early Discoverer missions faced numerous technical challenges. Discoverer Zero, launched on January 21, 1959, suffered a catastrophic failure when the Agena ullage and separation rockets ignited on the pad while the vehicle was being fueled, preventing launch. This marked the start of a pattern of technical difficulties in the program's early days.

Discoverer 1, launched on February 28, 1959, became the first of its kind to achieve orbit. However, its mission ended prematurely when it decayed on March 17, 1959. The next flight, Discoverer 2, launched on April 13, 1959, made history by becoming the first satellite to feature three-axis stabilization, but it ultimately failed to recover its capsule, marking a significant challenge for future recovery missions.

Several subsequent missions also met unfortunate ends. Discoverer 3, launched on June 3, 1959, succumbed to a guidance failure, causing the vehicle to fall into the Pacific Ocean. Discoverer 4, launched on June 25, 1959, failed due to insufficient Agena engine thrust, resulting in another Pacific Ocean crash. Similar problems plagued Discoverer 5 (August 13, 1959) and Discoverer 6 (August 19, 1959), which failed due to power supply issues and retro-rocket malfunctions, respectively.

By the end of 1959, Discoverer 7 (November 7, 1959) and Discoverer 8 (November 20, 1959) both failed due to satellite tumbles in orbit and eccentric orbits, rendering recovery impossible. Discoverer 9 (February 4, 1960) met a premature end when the Agena was accidentally damaged during on-pad servicing, while Discoverer 10 (February 19, 1960) suffered a control failure, leading to its destruction by the Range Safety Officer (RSO).

A significant milestone was finally achieved with Discoverer 13, launched on August 10, 1960. This mission successfully tested the capsule recovery system, marking the first successful capsule recovery from space. This breakthrough paved the way for future intelligence-gathering efforts from space. Just days later, on August 18, 1960, Discoverer 14 became the first mission to recover intelligence imagery from space, utilizing the KH-1 camera system. The successful recovery of imagery signaled a significant leap in reconnaissance capabilities during the Cold War.

Despite this success, challenges persisted. Discoverer 15 (September 13, 1960) succeeded in attaining orbit but failed to recover its capsule, which sank prior to retrieval. Discoverer 16 (October 26, 1960) encountered an Agena failure, and Discoverer 17 (November 12, 1960) failed due to a film malfunction before the camera could even operate.

However, the program continued to evolve. Discoverer 18, launched on December 7, 1960, became the first successful mission to use the KH-2 camera system, improving the quality of images captured from space. Discoverer 19, launched on December 20, 1960, was a test of the Missile Defense Alarm System, pushing forward the dual-use of reconnaissance technology for both space surveillance and missile detection.

By 1961, the Discoverer series reached new heights with the introduction of the KH-5 camera system, offering better resolution and intelligence-gathering capabilities. However, technical failures still plagued missions like Discoverer 22 (March 30, 1961), which experienced a control malfunction and failed to achieve orbit, and Discoverer 25 (June 16, 1961), where streaks throughout the recovered film limited the mission's success.

The introduction of the KH-3 camera system during Discoverer 29 (August 30, 1961) and subsequent missions signaled an upgrade in image quality, despite early difficulties with focus. Discoverer 32 (October 13, 1961) achieved recovery on orbit 18, though much of the film remained out of focus.

The series continued to encounter varying levels of success and failure throughout the early 1960s. Discoverer 33 (October 23, 1961) failed to achieve orbit, while Discoverer 36 (December 12, 1961) carried additional payloads like the first amateur radio satellite, OSCAR 1, into orbit alongside its reconnaissance mission.

By 1962, the Discoverer program began transitioning into more advanced satellite programs under the Corona program's umbrella. The FTV series introduced further improvements to reconnaissance satellites, culminating in better image quality and more reliable recovery mechanisms, despite continued challenges with issues such as corona static on film and equipment malfunctions.

The Discoverer series laid the foundation for modern satellite reconnaissance. Its early failures and eventual successes helped pave the way for the U.S. to gather critical intelligence during the Cold War, advancing the country's capabilities in space technology and defense. By the end of the series in 1972, significant strides had been made in space-based imaging, positioning the United States as a leader in reconnaissance from space.

Explorer 6

First photograph of Earth from orbit

On August 7, 1959, NASA launched Explorer 6, also known as S-2, a pioneering satellite designed to study various aspects of Earth's upper atmosphere and cosmic radiation. Launched at 14:24:20 GMT aboard a Thor-Able rocket from Cape Canaveral, Florida, Explorer 6 marked a significant step in the United States' early space exploration efforts. The satellite's primary scientific objectives were to investigate trapped radiation, cosmic rays, geomagnetism, radio propagation, micrometeorite flux, and cloud cover photography. It also carried a television scanner intended to capture images of Earth's cloud cover—a precursor to modern Earth-observation satellites.

Experiments and Discoveries

Explorer 6's most notable achievement occurred on August 14, 1959, when it transmitted the first photograph of Earth from space. Taken from an altitude of about 27,000 kilometers (17,000 miles), the image depicted the north-central Pacific Ocean, showcasing sunlit clouds over the region. Although rudimentary by modern standards, the photograph symbolized humanity's first glimpse of our planet from the vantage point of space.

Explorer 6 carried a suite of scientific instruments to conduct experiments that broadened understanding of space and Earth's environment:

Beacon (108 and 378 MHz): This experiment measured the electron density around the satellite using two coherent transmitters. Observations were conducted for up to 70 minutes during eight orbital passes. However, a strong magnetic storm and severe signal fading complicated data interpretation. The 378 MHz beacon failed early, limiting the experiment's duration to only 11 days.

Fluxgate Magnetometer: Designed to measure the magnetic field parallel to the satellite's spin axis, the fluxgate magnetometer, in combination with a search-coil magnetometer, was intended to provide a comprehensive understanding of the ambient magnetic field. Unfortunately, due to disturbances within the vehicle, the fluxgate magnetometer became saturated and failed to return useful data.

Ion Chamber and Geiger-Müller Counter: These instruments measured the flux of high-energy electrons and protons. The ion chamber operated normally until August 25, 1959, while the Geiger-Müller tube continued functioning until October 6, 1959. These measurements helped scientists study the intensity and distribution of cosmic radiation and trapped particles in Earth's magnetosphere.

Micrometeorite Detector: This instrument, designed to record the momentum and frequency of micrometeorites, detected pulses but ultimately failed to return significant scientific data.

Proportional Counter Telescope: Explorer 6 carried a dual proportional counter telescope to measure high-energy protons and electrons in space. The instrument allowed scientists to differentiate between high-energy radiation sources, such as X-rays and protons. Data from the telescopes provided valuable insights into the properties of cosmic radiation in interplanetary space.

Explore 6 paddles up

Scintillation Counter: This experiment measured electrons in Earth's radiation belts. Although the instrument worked briefly, its performance was hampered by the failure of the satellite's transmitter on September 11, 1959.

Search-Coil Magnetometer: Similar to the one flown on Pioneer 1, this magnetometer was intended to measure Earth's and interplanetary magnetic fields. Due to the satellite's low apogee, no interplanetary or lunar magnetic fields were detected. The instrument transmitted data until early October 1959, when the telemetry signal was lost.

TV Optical Scanner: Explorer 6's TV scanner, based on earlier designs used on Pioneer 2, was an innovative device to capture low-resolution images of Earth's cloud cover. Although the satellite succeeded in taking the first photograph of Earth from space, the image quality was poor due to malfunctioning circuits and incorrect spacecraft orientation. As a result, the final image was heavily distorted and incomplete, with useful data ceasing by August 25, 1959.

Very Low Frequency (VLF) Receiver (15.5 kHz): This experiment was designed to study Whistler mode propagation and ionospheric noise. Although data were recorded, technical issues reduced the receiver's sensitivity, limiting the experiment's success.

Explorer 6 was spin-stabilized, rotating at 2.8 revolutions per second, with solar panels mounted near its equator to recharge the satellite's batteries. However, only three of the four solar paddles fully deployed, reducing the satellite's power output to 63% of its intended capacity. This power deficiency affected the quality of the transmitted data, especially near the satellite's apogee.

On September 11, 1959, one of the satellite's Very High Frequency (VHF) transmitters failed, and by October 6, contact with Explorer 6 was lost as its solar cell charging current dropped below operational levels. In total, 827 hours of analog and 23 hours of digital data were collected during the mission.

Explorer 6's contributions extended beyond its scientific experiments. On October 13, 1959, it became the target of the Bold Orion anti-satellite (ASAT) test. The missile passed within 6.4 kilometers (4 miles) of the satellite, marking an early milestone in the development of anti-satellite technology during the Cold War.

Explorer 6's orbit finally decayed on July 1, 1961, ending its journey in space. Despite its challenges and technical limitations, the satellite significantly contributed to the early understanding of space environments. It marked a crucial step forward in satellite technology, Earth observation, and space-based scientific research. Its legacy lives on as a pioneer in the study of space, setting the stage for more advanced missions in the years that followed.

Luna 2

First hard landing on another celestial body (the Moon)

Luna 2, originally named the Second Soviet Cosmic Rocket and often called "Lunik 2" in contemporary media, marked a significant milestone in the Space Race. As part of the Soviet Union's Luna program, it became the first spacecraft to reach the surface of the Moon and the first human-made object to make contact with another celestial body. This historic achievement occurred on September 13, 1959, when Luna 2 successfully impacted the Moon, a feat that had eluded previous space missions. Its success not only boosted Soviet prestige but also intensified the competition between the United States and the Soviet Union in their pursuit of space dominance.

Luna 2 was the sixth spacecraft in the Soviet Luna program and was designated E-1 No.7. It followed the partially successful Luna 1 mission, which had been launched on January 2, 1959. Luna 1 missed the Moon by 5,965 kilometers, but it provided valuable insights for future missions. Between Luna 1 and Luna 2, an additional launch attempt in June 1959 ended in failure, but the Soviets remained determined to achieve lunar impact.

Designed as a spherical probe with antennas and scientific instruments, Luna 2 closely resembled its predecessor, Luna 1. It carried a suite of instruments designed to measure radiation, magnetic fields, and solar wind. Its primary scientific goal was to search for magnetic and radiation fields around the Moon, similar to the Van Allen radiation belts discovered around Earth. Weighing 390.2 kilograms, the spacecraft lacked propulsion systems of its own, relying entirely on the launch vehicle to deliver it on a direct trajectory to the Moon.

Luna 2 was launched on September 12, 1959, atop a Luna 8K72 rocket from the Baikonur Cosmodrome. After an initial launch attempt three days earlier had been aborted due to engine failure, the successful liftoff on September 12 sent Luna 2 on a path to make history. The spacecraft began transmitting telemetry data almost immediately after launch, allowing Soviet scientists to track its progress.

Once the spacecraft achieved escape velocity, it followed a direct route to the Moon. In addition to radio transmissions, Luna 2 released a cloud of sodium vapor into space, which expanded to 650 kilometers in diameter. This glowing cloud, visible from observatories across the Soviet Union, provided a visual marker for tracking the spacecraft's journey. The data from these observations allowed scientists to calculate that Luna 2 would impact the Moon on September 13.

Luna 2 impacted the Moon at 00:02:24 Moscow Time on September 14, 1959. It struck the lunar surface near the craters Aristides, Archimedes, and Autolycus, located in the region east of Mare Imbrium. The spacecraft impacted at a speed of approximately 3.3 kilometers per second, marking the first time in human history that an object created on Earth had made contact with another celestial body.

To commemorate the achievement, Luna 2 carried two pennants engraved with the emblem of the Soviet Union and the date of the mission. These pennants were designed to scatter pentagonal titanium shields across the lunar surface upon impact. This symbolic gesture served as a powerful demonstration of Soviet prowess during the Cold War space race.

Luna 2's scientific payload included several instruments, such as a triaxial fluxgate magnetometer, scintillation counters, Geiger counters, and ion traps. These instruments were designed to measure various aspects of the space environment as the spacecraft traveled to the Moon.

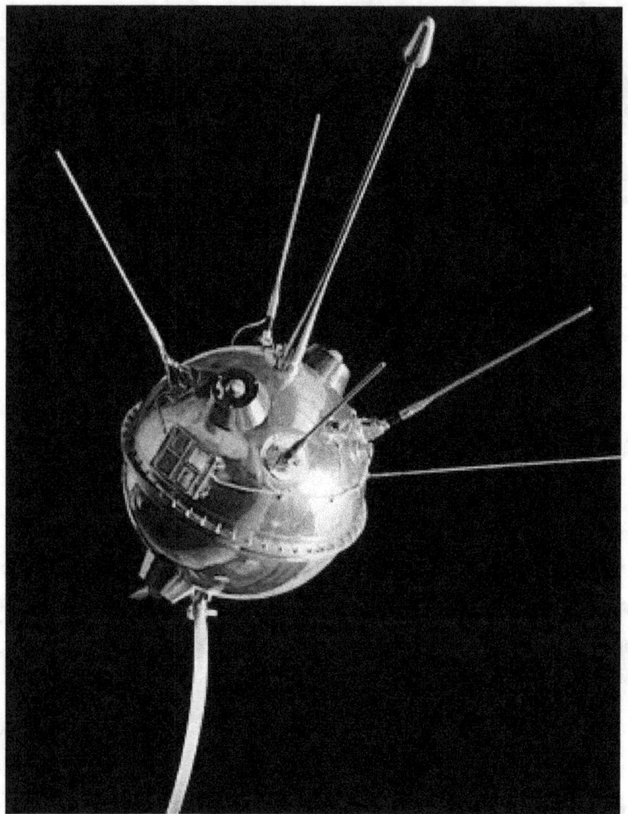

The mission yielded valuable data, including the absence of a magnetic field or radiation belts around the Moon. As Luna 2 approached the lunar surface, it detected no significant magnetic field within 55 kilometers of the Moon, confirming the Moon's lack of a magnetic shield similar to Earth's Van Allen belts. The probe also conducted the first direct measurement of solar wind flux outside Earth's magnetosphere, providing insights into the behavior of charged particles in space.

Although Luna 2's mission was brief, its scientific results were significant. The probe's data confirmed earlier measurements from Luna 1 and advanced humanity's understanding of the space environment between Earth and the Moon.

The Impact on the Space Race

Luna 2's success was a monumental achievement for the Soviet Union. The mission had far-reaching political and cultural implications, particularly in the context of the Cold War. At the time, Americans were optimistic about their progress in the space race, believing that while the Soviets had larger rockets, the United States had superior guidance systems. Luna 2's impact on the Moon shattered this confidence and highlighted the Soviet Union's technological capabilities.

Soviet Premier Nikita Khrushchev, visiting the United States shortly after the mission, presented President Dwight D. Eisenhower with a replica of the Soviet pennant left on the lunar surface. This gesture was both a diplomatic move and a demonstration of Soviet superiority in space exploration at the time.

Luna 2's technological advancements did not go unnoticed by the West. In 1959, as part of a covert operation, the CIA gained access to a fully operational Luna 2 display at an exhibition of Soviet economic achievements. Expecting the display to be a replica, CIA officers were surprised to find that the spacecraft was an original system. Over 24 hours, they disassembled and photographed the spacecraft, gathering valuable intelligence about Soviet space technology.

This intelligence operation, which remained undisclosed until the declassification of CIA documents in 2019, highlighted the high stakes of the space race, where both superpowers closely monitored technological breakthroughs.

Luna 2 was not only a victory for the Soviet Union but also set the stage for future space missions. It was the first in a series of intentional lunar impact missions, demonstrating that hard landings could provide valuable scientific data. NASA later adopted this approach during its Ranger program, which also involved intentional crashes on the lunar surface.

The legacy of Luna 2's hard impact missions persisted into the modern era, as controlled impacts were later used to test for the presence of ice in shadowed lunar craters. By analyzing debris ejected from these impacts, scientists gained insights into the Moon's composition and potential resources, such as water ice, vital for future lunar exploration.

Luna 3

First three-axis stabilised spacecraft
First photos of far side of the Moon, covering 70% of the surface invisible from Earth
First automated on board development of photographic film and conversion to radio signals
First gravity assist ('sling shot'), returning the spacecraft to Earth to retrieve the photos

In 1959, as part of the Soviet Union's Luna program, the Luna 3 spacecraft, also known as E-2A No. 1, became the first mission to capture photographs of the Moon's far side. This achievement was groundbreaking, as it provided humanity with its first glimpse of the side of the Moon that faces away from Earth—a side never before seen by human eyes. Luna 3 was the third Soviet spacecraft sent to the vicinity of the Moon, following the successful Luna 1 and Luna 2 missions.

The spacecraft, a cylindrical canister with hemispherical ends, measured 130 centimeters in length and had a diameter of approximately 95 centimeters. The design was simple yet efficient: a pressurized, hermetically sealed canister equipped with solar cells for power and several scientific instruments mounted externally, including micrometeoroid and cosmic ray detectors. Its internal equipment included cameras, a photographic film processing system, and gyroscopic units for attitude control. The spacecraft was spin-stabilized for most of the mission, but when it came time to photograph the Moon, Luna 3 used a three-axis attitude control system to achieve the necessary precision for imaging.

Launched on a Luna 8K72 rocket, the spacecraft embarked on its mission by passing over the Earth's North Pole and heading toward the Moon. After overcoming early technical challenges—including weak signal strength and rising internal temperatures—Luna 3 approached within 6,200 kilometers of the lunar surface near the Moon's south pole. On October 7, 1959, as it passed over the far side of the Moon, the spacecraft's photocell detected sunlight illuminating the surface, triggering the start of the photography sequence. Over the next 40 minutes, Luna 3 captured 29 images, covering 70% of the Moon's far side from distances between 63,500 and 66,700 kilometers.

These images revealed a landscape markedly different from the familiar near side of the Moon. While the near side is characterized by vast, dark plains known as maria, the far side appeared mountainous with few of these low-lying regions. Only two dark regions were initially identified and named: Mare Moscoviense (Sea of Moscow) and Mare Desiderii (Sea of Desire). However, further analysis revealed that Mare Desiderii was actually composed of smaller features, including Mare Ingenii (Sea of Cleverness) and other dark craters.

The images from Luna 3 caused a sensation when published around the world. Despite some transmission difficulties due to the spacecraft's weak signal, 17 of the 29 photographs were successfully transmitted back to Earth. Six of these images were eventually published, and a preliminary "Atlas of the Far Side of the Moon" was created based on the data. The photos were not only significant for their historical value but also for the scientific insight they provided into the geological differences between the Moon's two hemispheres.

One of the mission's most notable accomplishments was its use of a gravity assist maneuver—the first of its kind in space exploration. As Luna 3 passed behind the Moon, the gravitational pull of the Moon altered the spacecraft's trajectory, effectively slingshotting it back toward Earth. This maneuver, which also changed the spacecraft's orbital plane, allowed Luna 3 to return to Soviet ground stations for data transmission. The success of this gravity assist relied on the pioneering research of Soviet mathematician Mstislav Keldysh and his team at the Steklov Institute of Mathematics.

The first view returned by Luna 3 showed the far side of the Moon was very different from the near side, most noticeably in its lack of lunar maria (the dark areas)

The Luna 3 mission was a triumph of Soviet engineering and scientific ingenuity. The spacecraft's imaging system, Yenisey-2, was developed at the Leningrad Scientific Research Institute for Television and consisted of dual-lens cameras, an automatic film processing unit, and a scanner. The system used 35mm temperature- and radiation-resistant film, a critical feature for the harsh space conditions. The cameras had focal lengths of 200 mm and 500 mm, allowing them to capture both wide-angle images of the Moon's surface and detailed close-ups.

Once the images were captured, the film was developed and scanned aboard the spacecraft. The data was transmitted back to Earth using frequency-modulated analog video signals received by Crimea and Kamchatka tracking stations. Despite the technological limitations of the time, the images were remarkably clear and provided a wealth of new information about the Moon's far side.

Discoverer 13

First satellite recovered intact from orbit

On August 10, 1960, at precisely 20:37:54 GMT, the United States launched Discoverer 13, an optical reconnaissance satellite, from Vandenberg Air Force Base. This satellite was part of the Corona program, which sought to replace the U-2 spy plane with space-based reconnaissance, focusing on the Sino-Soviet Bloc to monitor the production and deployment of Soviet missiles and long-range bombers. As the final test flight of the Corona Keyhole 1 (KH-1) spy satellite series, Discoverer 13 marked a significant turning point in American space and intelligence efforts, becoming the first fully successful mission in the Discoverer series.

The "Discoverer" name served as the civilian cover for the highly classified Corona project, managed by the Department of Defense's Advanced Research Projects Agency (ARPA) and the U.S. Air Force. The goal of the Corona satellites was not only to gather intelligence on Soviet military capabilities but also to produce maps and charts for U.S. defense and mapping agencies. The early iterations of the Corona satellites, designated Keyhole 1 (KH-1), relied on the Agena-A upper stage, which housed the satellite and provided crucial attitude control in orbit.

Equipped with a vertical-looking panoramic camera built by Fairchild Camera and Instrument, the KH-1 satellites could capture sweeping images of the Earth's surface with a ground resolution of 12.9 meters. The photographic film from these missions was stored in a Satellite Return Vehicle (SRV) developed by General Electric, which would reenter the Earth's atmosphere at the end of the mission, its contents to be recovered mid-air by specially equipped aircraft.

The Discoverer program experienced a series of setbacks before its first success. Operational attempts followed initial test flights in 1959, but the missions suffered from technical malfunctions and failures. By 1960, the program still had not achieved a fully successful flight. Discoverer 11, launched on April 15, 1960, came close—its camera worked properly for the first time—but a spin motor malfunction caused a failure during reentry. Discoverer 12, launched on June 29, 1960, was intended to diagnose the recurring problems but was lost when its Agena booster malfunctioned shortly after launch.

Discoverer 13, launched on August 10, 1960, was nearly identical to its predecessor, Discoverer 12, with several improvements to ensure mission success. Both were equipped with a new gas motor for spin stabilization to replace the faulty system that had doomed Discoverer 11. While previous satellites carried surveillance cameras, Discoverer 13's payload consisted of additional telemetry instruments, a doppler beacon, and an American flag as part of the mission's symbolic and public diplomacy goals.

On August 11, after 17 orbits, Discoverer 13's reentry capsule was commanded to return to Earth. The satellite's Agena upper stage pitched down 60 degrees, and the SRV was ejected using small springs. A newly developed spin engine stabilized the capsule before its retro-rocket fired, reducing its velocity by 400 meters per second. As the SRV descended, a series of parachutes deployed, slowing the capsule and allowing for recovery. Though the plan was to catch the capsule mid-air, the recovery aircraft missed its mark, and the SRV splashed down in the Pacific Ocean, 610 kilometers northwest of Honolulu.

The recovery ship, Haiti Victory, dispatched a helicopter that deployed divers to secure the capsule for retrieval. The SRV was then transported to Pearl Harbor in Hawaii. Four days later, on August 15, 1960, the American flag carried aboard the satellite was presented to President Dwight D. Eisenhower, a moment that highlighted the mission's success and masked the true intelligence purpose of the Corona program.

The success of Discoverer 13 was a monumental achievement, as it validated the design and operational capabilities of the Corona series. Eight days later, on August 18, 1960, the United States launched Discoverer 14, the first fully operational Corona reconnaissance satellite. Unlike its predecessors, Discoverer 14 successfully returned photographic intelligence of the Soviet Union, marking the beginning of an era of space-based reconnaissance.

The Corona program ultimately spanned 145 flights in eight satellite series, with the final mission launched on May 25, 1972. The Corona project remained classified until its declassification in 1995, when the U.S. government formally acknowledged the existence of its reconnaissance satellite programs. The legacy of Corona and the Discoverer series paved the way for modern satellite intelligence, significantly contributing to national security during the Cold War.

In a symbolic footnote to the mission, Lockheed employees who worked on the Corona project celebrated Discoverer 13's success with a party, throwing the program's manager, James Plummer, into a hotel pool in East Palo Alto, California—an exuberant gesture that marked the triumph of years of hard work and perseverance.

Discoverer 14

First spy photography from space
First aerial recovery of an object (the film) returning from Earth orbit

The Discoverer 14 satellite, also designated Corona 9009, marked a significant moment in the history of space-based reconnaissance. Launched on August 18, 1960, as part of the U.S. Corona program, it was the first satellite to successfully recover usable photographic images of the Soviet Union from orbit. This breakthrough, managed by the Advanced Research Projects Agency (ARPA) of the Department of Defense in conjunction with the United States Air Force, signaled a significant advancement in surveillance technology during the Cold War.

The Discoverer 14 satellite was launched atop a Thor-Agena A booster from Vandenberg Air Force Base. The satellite's camera, manufactured by Fairchild Camera and Instrument, featured an impressive 61-centimeter focal length and a ground resolution of 42 feet, allowing it to capture detailed images from space. Film used in the satellite was stored in a General Electric Satellite Return Vehicle (SRV), which would return to Earth, where it was recovered mid-air by specially outfitted aircraft.

The first Corona photo

On August 18, 1960, Discoverer 14 was launched into a polar orbit, a crucial trajectory for capturing images of the Soviet Union. Although the satellite initially experienced stability issues during its first few orbits, it eventually stabilized and began normal operations. On its 17th pass around the Earth, Discoverer 14's SRV, containing 20 pounds of film, was ejected over Alaska.

What followed was a groundbreaking moment in space history: as the SRV reentered Earth's atmosphere, it deployed a parachute. A C-119 recovery aircraft from the 6593rd Test Squadron, piloted by Captain Harold E. Mitchell, successfully intercepted and recovered the capsule mid-air, approximately 360 miles southwest of Honolulu. This was the first successful film recovery from an orbiting satellite and the first mid-air recovery of an object returning from space. Captain Mitchell was awarded the Bronze Star, and his crew received Air Medals for their exceptional achievement. The success of Discoverer 14 marked the dawn of a new era in reconnaissance. The satellite's images provided more extensive photographic coverage of the Soviet Union than had been gathered in all previous U-2 missions combined. These images were vital for U.S. intelligence, offering unprecedented insight into Soviet military capabilities during escalating Cold War tensions.

The Corona program would eventually span 145 missions over 12 years, concluding in May 1972.

Korabl-Sputnik 2 (aka Sputnik 5)

First animals and plants returned alive from space (the dogs Belka and Strelka)
First capsule recovered from orbit

Korabl-Sputnik 2, also known as Sputnik 5 in the West, was a significant Soviet space mission and the third test flight of the Vostok spacecraft. Launched on August 19, 1960, it marked a major milestone in the Space Race by becoming the first spaceflight to send animals into orbit and return them safely to Earth. Among the crew were two Soviet space dogs, Belka and Strelka, accompanied by forty mice, two rats, and a variety of plants. This mission paved the way for future human spaceflight, culminating in Yuri Gagarin's historic Vostok 1 flight less than eight months later.

Korabl-Sputnik 2 was the second attempt by the Soviet Union to launch a Vostok capsule carrying animals into orbit. The first attempt, launched on July 28, 1960, ended in tragedy. The spacecraft, carrying two dogs named Bars (also known as Chaika) and Lisichka, was lost when a fire broke out in the Blok G strap-on booster. The booster disintegrated 19 seconds after liftoff, scattering its remains over the steppe. Flight controllers attempted to separate the descent module and deploy the parachutes, but due to the low altitude of the separation, the parachutes failed to deploy fully. Tragically, both dogs were killed on impact.

This failure exposed issues with the RD-107 engines, a persistent problem in earlier R-7 launches due to high-frequency pressure oscillations. Although the RD-107 had undergone revisions to fix these issues, the booster used for this launch was fitted with an older engine version. In response to this disaster, Soviet engineers accelerated the development of an ejector seat for cosmonauts to ensure a means of escape in the event of a launch failure. This development would prove critical for future crewed missions.

Korabl-Sputnik 2's preparations were also complicated by a two-day delay caused by the need to replace a faulty liquid oxygen valve. Despite these setbacks, the mission became a resounding success.

On August 19, 1960, at 08:44:06 UTC, Korabl-Sputnik 2 lifted off atop a Vostok-L carrier rocket from the Baikonur Cosmodrome. Radio stations in West Germany and Sweden confirmed signals from the spacecraft almost immediately after launch, providing early indications that the mission was proceeding smoothly.

The spacecraft's biological payload, consisting of Belka, Strelka, mice, rats, and plant specimens, was monitored closely via onboard telemetry. A television camera aboard the spacecraft captured images of the two dogs, allowing ground controllers to observe their behavior in space. However, during the fourth orbit, telemetry revealed that Belka experienced seizures and vomiting—an indication that prolonged exposure to weightlessness could present challenges for living organisms. As a result of this observation, Soviet mission planners decided to limit the first human orbital flight, Vostok 1, to three orbits to minimize potential health risks.

Mercury-Redstone 2

First great ape or Hominidae in space, Ham, a chimpanzee

Mercury-Redstone 2 (MR-2) was a significant mission in the early stages of America's space exploration efforts, designed as a test flight of the Mercury-Redstone Launch Vehicle in preparation for the first crewed mission under Project Mercury. The launch took place on January 31, 1961, at 16:55 UTC from Launch Complex 5 at Cape Canaveral, Florida. The spacecraft, Mercury Number 5, carried a chimpanzee named Ham on a suborbital flight, marking a significant moment in the history of human spaceflight. This mission served as a final step before sending a human into space.

Before MR-2, the Mercury-Redstone program experienced challenges. The preceding mission, Mercury-Redstone 1A (MR-1A), followed a too steep trajectory, subjecting the spacecraft to higher-than-anticipated acceleration forces. The MR-1A mission reached an apogee of approximately 130 miles (210 kilometers) and landed 235 miles (378 kilometers) downrange, but its flight dynamics proved unsuitable for human passengers. Learning from this, the MR-2 mission was designed with a flatter trajectory, aiming for an apogee of 115 miles (185 kilometers) and a range of 290 miles (470 kilometers).

In early January 1961, six chimpanzees—four females and two males—were selected and trained for the mission at Holloman Air Force Base in New Mexico. These animals were chosen for their ability to perform tasks under stress, with Ham eventually selected as the primary candidate for the flight. Ham, a 3-year-old male chimpanzee originally from Cameroon, had been purchased by the U.S. Air Force in 1959 and named in honor of the Holloman Aerospace Medical Center. His backup for the mission was a female chimpanzee named Minnie.

The mission introduced six new systems to the Mercury spacecraft, including an environmental control system, attitude stabilization control, live retrorockets, voice communications, a closed-loop abort sensing system, and a pneumatic landing bag. On January 2, 1961, the chimpanzees were transferred to Cape Canaveral, where they underwent three weeks of intensive training in Mercury simulators. Ham was chosen as the flight's primary astronaut the day before the mission, with Minnie as the backup.

Ham was secured in the Mercury spacecraft at 12:53 UTC on January 31, 1961, but the launch was delayed due to technical issues, including a malfunctioning inverter. After nearly four hours of delay, the MR-2 mission finally lifted off at 16:55 UTC.

Shortly after launch, it became apparent that the flight path angle was steeper than planned, a concerning development that worsened as the spacecraft ascended. Two minutes into the flight, the mission's systems detected that the spacecraft's acceleration was significantly higher than expected, reaching a maximum of 17 g. At 2 minutes and 17 seconds, the Redstone rocket's liquid oxygen supply was depleted, and the spacecraft's launch escape system activated, triggering an abort sequence. This early termination caused the spacecraft to overshoot its intended landing area by 130 miles (210 kilometers) and reach an apogee of 157 miles (253 kilometers) instead of the planned 115 miles.

A further complication occurred when cabin pressure dropped from 5.5 psi to just 1 psi (38 to 7 kPa), caused by a malfunctioning snorkel valve. However, Ham was unaffected, thanks to his personal spacesuit, which maintained a stable environment with the temperature well within safe limits of 60 to 80 degrees Fahrenheit (16 to 26 °C).

The spacecraft reached a velocity of 5,140 mph (8,270 km/h), well above the planned 4,400 mph (7,100 km/h). As a result, Ham experienced weightlessness for 6.6 minutes instead of the planned 4.9 minutes. Despite these anomalies, Ham performed exceptionally well, completing his assigned task of pulling levers about 50 times during the flight.

After a flight lasting 16 minutes and 39 seconds, the capsule splashed down in the Atlantic Ocean, 422 miles (679 kilometers) downrange from Cape Canaveral. The spacecraft initially landed upright but soon began taking on water after capsizing due to damage inflicted on the capsule during reentry. The beryllium heat shield had punctured the titanium pressure bulkhead, allowing seawater to enter the capsule through an open snorkel valve.

Rescue forces were initially unable to locate the spacecraft, as it had splashed down out of sight of recovery ships. After 27 minutes, a search plane spotted the capsule, and helicopters were dispatched to retrieve it. When recovery teams finally reached the spacecraft, they found it submerged and filled with approximately 800 pounds (360 kilograms) of seawater. Nevertheless, Ham was in good condition, eagerly accepting an apple and half an orange after being extracted from the capsule.

Although Ham's performance was exemplary, the malfunctions during the flight raised concerns about the Mercury-Redstone system's readiness for a human astronaut. As a result, NASA postponed the first crewed mission, initially scheduled for MR-3, pending further tests.

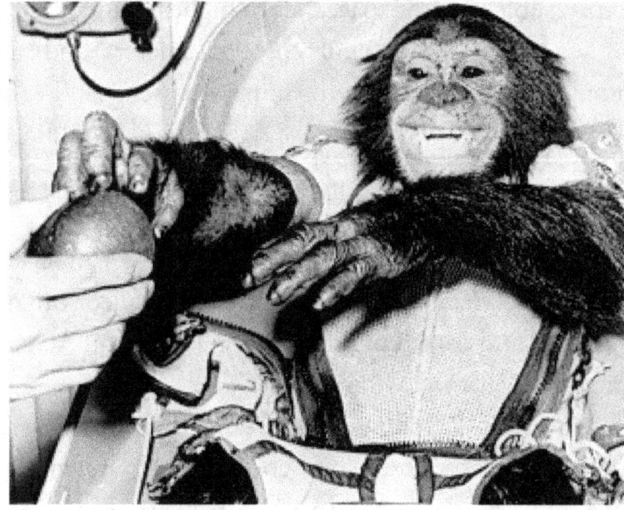

Ham rewarded with an apple

Chimpanzee "Ham" in space suit is fitted into the couch
of the Mercury-Redstone 2 capsule #5 prior to its test
flight on January 31, 1961.

Ham, having secured his place in spaceflight history as the first great ape to journey into space, was transferred to the National Zoo in Washington, D.C., where he lived for 17 years. In 1981, he was moved to a zoo in North Carolina and spent his remaining years among a colony of chimpanzees before passing away in 1983 at 26. Ham is buried at the New Mexico Museum of Space History in Alamogordo, New Mexico.

His backup, Minnie, also played a significant role in space history, though she never flew a mission. She became part of an Air Force breeding program and was the last surviving "astro-chimp," living to 41 before passing away in 1998.

Timeline of Events

Liftoff: MR-2 launched at 16:55 UTC.

Flight Anomaly: By T+1:00, computers reported a steepening flight path angle.

Max Q: At T+1:24, maximum dynamic pressure occurred, reaching 575 lbf/ft^2.

BECO and Abort: At T+2:17, booster engine cutoff occurred three seconds early, triggering the abort sequence.

Capsule Separation: The capsule separated from the booster at T+2:17.

Cabin Pressure Malfunction: At T+2:18, cabin pressure dropped due to a malfunctioning snorkel valve.

Splashdown: The capsule splashed down at T+16:39, 422 miles downrange from the launch site.

The Mercury-Redstone 2 mission, while marked by numerous technical challenges, proved invaluable in ensuring the safety of future human spaceflights, paving the way for Alan Shepard's historic journey aboard Mercury-Redstone 3 just a few months later.

Venera 1

First launch from Earth orbit of upper stage into a heliocentric orbit
First mid-course corrections
First spin-stabilisation
First planetary flyby (Venus), although contact was lost

In February 1961, the Soviet Union launched Venera 1, marking a significant milestone in the Space Race as the first spacecraft to perform an interplanetary flight and fly past Venus. Known in the West as Sputnik 8, Venera 1 was part of the ambitious Venera program aimed at exploring Venus, our nearest planetary neighbor. Although radio contact was lost before its scheduled flyby on May 19, 1961, Venera 1 set the stage for future interplanetary missions and demonstrated the Soviet Union's growing expertise in space exploration.

Weighing 643.5 kilograms (1,419 pounds), Venera 1 had a cylindrical body measuring 1.05 meters (3 feet 5 inches) in diameter, topped with a dome that brought the total height to 2.035 meters (6 feet 8.1 inches). This pressurized probe was filled with dry nitrogen and maintained thermal stability using internal fans to distribute heat evenly. Two solar panels extended from the sides of the cylindrical body, providing energy to charge its silver-zinc batteries.

Venera 1 was equipped with an array of sophisticated antenna systems for communication. A 2-meter (6-foot 7-inch) parabolic wire-mesh antenna was designed to transmit data from Venus back to Earth at a frequency of 922.8 MHz, while a 2.4-meter (7-foot 10-inch) antenna boom was used to send short-wave signals during the early phases of the mission. Routine telemetry was handled by semidirectional quadrupole antennas mounted on the solar panels.

Venera 1 was outfitted with a suite of scientific instruments, including a flux-gate magnetometer attached to the antenna boom for measuring magnetic fields, ion traps to detect solar wind, and micrometeorite detectors. The probe carried Geiger counters and a sodium iodide scintillator to measure cosmic radiation. Additionally, an experiment on one of the solar panels measured the temperature of different coatings, and infrared or ultraviolet radiometers may have been included to gather more data. The spacecraft's mid-course corrections were made possible by the KDU-414 engine, located in the dome, which would also have controlled temperature through motorized thermal shutters.

Venera 1 introduced several technological innovations. During most of its mission, the spacecraft was spin-stabilized, a common practice at the time to maintain a stable trajectory. However, it was the first spacecraft designed to perform mid-course corrections using three-axis stabilization. This advanced method relied on fixing the spacecraft's orientation with respect to both the Sun and the star Canopus. This capability would have allowed Venera 1 to correct its path as it approached Venus and switched to a mode of fixing its position on the Sun and Earth, employing a parabolic antenna to relay data back to Earth for the first time.

Spacecraft Venera 1 a mockup

Venera 1 was the second of two attempts to send a spacecraft to Venus in early 1961. Its sister probe, Venera-1VA No.1, had failed to leave Earth orbit. On February 12, 1961, at 00:34:36 GMT, Soviet engineers successfully launched Venera 1 aboard a Molniya carrier rocket from the Baikonur Cosmodrome. The spacecraft and its Blok-L upper stage were first placed into a low Earth orbit, where the upper stage ignited to propel the spacecraft into a heliocentric trajectory toward Venus. The 11D33 engine used for this mission phase was groundbreaking—it was the world's first staged-combustion-cycle rocket engine, and its ullage engine made it possible to restart a liquid-fuel rocket engine in space, a first in rocket technology.

Despite early successes, the mission was ultimately hampered by communication issues. Venera 1's initial telemetry sessions were successful, and the probe provided critical data on solar wind and cosmic rays. Data was transmitted during three telemetry sessions, measuring solar wind near Earth, at Earth's magnetopause, and on February 19, when the spacecraft had traveled 1.9 million kilometers (1.2 million miles) from Earth. This confirmed the discovery made by Luna 2 that solar wind—a stream of charged particles emitted by the Sun—was present throughout deep space. Venera 1's data verified this plasma's uniformity, marking an important discovery in space science.

However, contact was lost after this final transmission, and the next scheduled telemetry session failed. On May 19, 1961, Venera 1 passed within 100,000 kilometers (62,000 miles) of Venus, but due to the communication failure, no data from the flyby was returned. Despite the loss of contact, the British Jodrell Bank radio telescope possibly detected weak signals from the spacecraft in June, offering a faint hope of recovery.

Soviet engineers concluded that the likely cause of the failure was the overheating of a solar-direction sensor. This vital component helped the spacecraft maintain its orientation toward the Sun. Though the mission did not return scientific data from Venus, Venera 1's achievements in space technology and navigation were significant, and it paved the way for future successes in interplanetary exploration.

Although Venera 1 ultimately failed to fulfill its mission objectives, it represented a remarkable technological leap in the early years of the Space Race. As the first spacecraft to fly past another planet, it set the stage for more advanced missions in both the Soviet and American space programs. Venera 1's innovative use of mid-course corrections and its contributions to understanding solar wind have earned it a prominent place in the history of space exploration, laying the groundwork for the later successes of the Venera program and the broader Soviet efforts to explore the cosmos.

Chapter 3 - Pioneers of Spaceflight

Vostok 1

First human spaceflight mission (Yuri Gagarin)
First orbital flight of a manned vehicle

On April 12, 1961, the world was forever changed as the Soviet Union launched Vostok 1, marking the first human orbital spaceflight in history. This milestone, part of the Soviet Union's ambitious Vostok program, propelled Yuri Gagarin into the annals of history as the first person to travel into space and complete an orbit of Earth. The mission, which lasted 108 minutes from launch to landing, showcased Soviet technological prowess at the height of the Cold War, igniting the Space Race between the United States and the Soviet Union.

Part of the Vostok 1 instrument panel prominently
displaying the "Globus" navigation instrument

The origins of the Space Race can be traced back to 1957, when the Soviet Union launched the world's first artificial satellite, Sputnik 1. This triumph spurred both superpowers—locked in Cold War rivalry—to rapidly develop spaceflight capabilities, with each aiming to be the first to send a human into space. The Soviet Union's Vostok program competed directly with the United States' Project Mercury, both focused on developing the technology to achieve human spaceflight. Several uncrewed missions were conducted to test the Vostok spacecraft before Gagarin's flight. The final test flights, Korabl-Sputnik 4 and Korabl-Sputnik 5, were resounding successes, clearing the way for the historic crewed mission of Vostok 1.

Yuri Gagarin was selected as the primary pilot for Vostok 1, with Gherman Titov and Grigori Nelyubov as backups. On April 9, 1961, Gagarin and Titov were informed of the final decision. Gagarin, who had been a leading candidate for months, was overjoyed to be selected for the mission, while Titov, disappointed, would soon take part in Vostok 2. Gagarin's calm demeanor and readiness for the mission played a significant role in his selection despite the high levels of secrecy surrounding the entire operation.

On April 12, 1961, the Vostok 1 spacecraft was prepared for launch at the Baikonur Cosmodrome. Gagarin and his backup, Titov, received a final flight plan review and were transported to the launch pad. Gagarin, dressed in his space suit, entered the capsule at 07:10 local time, and the spacecraft's radio system was activated. The mission remained on schedule despite some last-minute technical issues with the spacecraft's hatch.

Model of the Vostok spacecraft with its upper stage,

Soviet engineers took every precaution to ensure Gagarin's safety. The spacecraft was equipped with 13 days' worth of provisions in case the retrorockets failed, allowing Gagarin to survive in orbit until natural orbital decay would bring him back to Earth. Gagarin's food was packed in specially designed metal tubes, and the letters "CCCP" were hand-painted on his helmet to ensure he wouldn't be mistaken for a foreign spy upon landing—a particular concern given recent Cold War tensions, including the U-2 incident.

At precisely 06:07 UTC, the Vostok-K rocket carrying Vostok 1 lifted off from Baikonur Cosmodrome. As the rocket ascended, Gagarin famously exclaimed, "Poyekhali!"—"Let's go!"—a phrase that would become a symbol of the Space Age. The ascent proceeded smoothly, with Gagarin reporting good visibility of Earth from space as the spacecraft moved into orbit. Just nine minutes after launch, Vostok 1 successfully entered orbit, confirming the Soviet Union's first human spaceflight achievement.

Time in Orbit

For nearly 90 minutes, Gagarin circled the Earth at altitudes ranging from 169 kilometers to 327 kilometers, making observations of the planet's surface. He communicated with ground control, reporting on the spacecraft's condition and his own well-being. Despite brief interruptions in communication, Gagarin remained calm and reported feeling fine throughout the flight. The spacecraft's automatic systems were responsible for controlling most aspects of the mission, including alignment for reentry, as it was uncertain how humans would react to the experience of weightlessness.

Gagarin with Korolev (right) before the flight

At 07:25 UTC, approximately 90 minutes after launch, the retrorockets fired over the west coast of Africa, initiating Vostok 1's reentry into Earth's atmosphere. However, a technical malfunction occurred when the service module failed to detach from the reentry capsule, causing the spacecraft to gyrate violently. Fortunately, the wires holding the two modules together burned away, allowing the capsule to reorient and continue a safe descent.

Gagarin experienced intense g-forces, peaking at around 8 g, as the spacecraft hurtled through the atmosphere. At an altitude of 7 kilometers, the hatch was released, and Gagarin ejected from the capsule, deploying his parachute shortly thereafter. Ten minutes later, at 08:05 UTC, Yuri Gagarin safely landed in a rural area near Engels, Saratov, where he was greeted by local villagers startled by his strange appearance. Ever modest, Gagarin reassured them, explaining that he was a Soviet citizen who had just returned from space.

The news of Gagarin's flight spread quickly around the world. Soviet citizens celebrated the event as a triumph of socialism and Soviet science, with mass demonstrations nationwide. Gagarin became an instant hero, receiving the title of Hero of the Soviet Union, the nation's highest honor. His achievement was a significant propaganda victory for the Soviet Union, demonstrating its technological superiority when the United States was still working toward its first manned spaceflight.

Internationally, reactions were mixed. While world leaders such as Jawaharlal Nehru of India praised the achievement as a victory for peace, many in the West viewed it through the lens of Cold War tensions. U.S. President John F. Kennedy congratulated the Soviet Union but acknowledged the need for the United States to accelerate its own space program. The flight of Vostok 1 spurred American efforts to catch up, ultimately leading to the Apollo program and the eventual U.S. moon landing in 1969.

Gagarin's words, "Poyekhali!" became immortalized in Soviet culture and the broader space community as a symbol of the dawn of human space exploration. His mission remains a landmark event in history, marking humanity's first steps into the vast unknown of space. April 12, the day of Gagarin's flight, is now celebrated internationally as the International Day of Human Space Flight, commemorating the courage and innovation that made the flight possible.

Freedom 7

First pilot-controlled space flight (Alan Shepard)

On May 5, 1961, the United States took a significant step in the Space Race with Mercury-Redstone 3, known as Freedom 7, the nation's first human spaceflight. Piloted by astronaut Alan Shepard, this mission marked a significant achievement in Project Mercury, which aimed to put an American astronaut into orbit and safely return them to Earth. Shepard's mission was a suborbital flight, lasting just over 15 minutes, with the primary objective of testing his ability to withstand the intense g-forces experienced during launch and re-entry.

Shepard's spacecraft, Freedom 7, symbolized personal and national pride. Continuing the tradition of pilots naming their crafts, Shepard's decision to name the capsule reflected this heritage and set a precedent for future astronauts in the Mercury program. The number "7" appended to the name was not merely a nod to NASA's "Mercury Seven" astronauts, as many believed, but referenced the McDonnell Model 7 spacecraft used in the mission. Shepard's spacecraft reached an altitude of 101.2 nautical miles (187.5 km) and traveled 263.1 nautical miles (487.3 km) downrange, launching from Cape Canaveral, Florida, near the Atlantic coast.

The mission's technical success was undeniable, though it came three weeks after the Soviet Union had sent Yuri Gagarin into orbit aboard Vostok 1, dampening the American public's excitement. Nonetheless, Shepard's flight showcased American ingenuity and courage in the face of fierce competition. National pride was rekindled when, in 2017, May 5 was designated National Astronaut Day to honor this landmark achievement.

Freedom 7's spacecraft, Mercury capsule #7, arrived at Cape Canaveral on December 9, 1960. However, extensive development and testing delayed the launch, pushing the tentative date from March 6 to early April. Safety concerns further postponed the mission. The earlier MR-2 test flight, carrying a chimpanzee named Ham, had experienced technical malfunctions, leading to the spacecraft traveling farther and faster than planned. This resulted in Ham enduring g-forces far greater than anticipated. These issues prompted NASA to add an additional test flight, MR-BD, on March 28, 1961. The success of this flight ensured that the crewed MR-3 mission could proceed with minimal further delay.

By early January 1961, Robert R. Gilruth, head of the Mercury program, had selected Alan Shepard as the primary pilot for MR-3, with John Glenn and Gus Grissom as backups. The crew was officially announced in February, though Shepard's name was kept confidential until shortly before the flight to allow flexibility in the event of any last-minute changes.

On the morning of May 5, 1961, Alan Shepard began his preflight preparations with a hearty breakfast of steak and eggs, a meal that would become an astronaut tradition. He entered the capsule at 5:15 AM, two hours before the scheduled launch. However, inclement weather and technical glitches caused several delays, leaving Shepard in the capsule for nearly three hours. As the countdown dragged on, Shepard, uncomfortable and needing to relieve himself, ultimately urinated in his suit after the medical sensors were turned off.

Finally, at 9:34 AM ET, Freedom 7 lifted off, with 45 million viewers watching on television. Shepard experienced a maximum acceleration of 6.3 g's just before the Redstone rocket engine shut down after two minutes and 22 seconds of flight. Upon separation from the rocket, the spacecraft's attitude control system successfully stabilized the capsule, which Shepard later manually tested. His skill in maneuvering the capsule proved that an astronaut could control a spacecraft, laying the groundwork for future missions.

Shepard conducted key tests during the flight, including operating the capsule's attitude control system and visually observing Earth through the spacecraft's periscope. He could easily identify major landmasses, such as Florida and the Bahamas, despite cloud cover obstructing parts of his view. Although the spacecraft had windows, the periscope remained the primary observational tool for this mission.

As Freedom 7 reached the apex of its trajectory, Shepard prepared for re-entry. He manually adjusted the spacecraft's pitch but faced confusion due to an error in the pitch indicator's reference settings, which had been updated without his knowledge. Despite this, the retrorockets fired successfully, beginning the spacecraft's descent back to Earth. Shepard switched to "fly-by-wire" mode, a semi-automatic system that gave him control over the re-entry while still relying on the spacecraft's automated systems for stabilization.

Astronaut Alan B. Shepard, Jr. sits in his *Freedom 7*
Mercury capsule, ready for launch. Just 23 days earlier,
Soviet cosmonaut Yuri Gagarin had become the first man
in space. That little race between Gagarin and me,
Shepard said, was really, really close. After several
delays and more than four hours in the capsule, Shepard
was ready to go, and he famously urged mission
controllers to fix your little problem and light this candle.

Freedom 7 re-entered Earth's atmosphere with g-forces peaking at 11.6 g's. The capsule descended rapidly, but the parachutes deployed as planned, and Shepard splashed down in the Atlantic Ocean, 263.1 nautical miles from Cape Canaveral. His spacecraft tipped slightly upon landing but quickly righted itself. Within minutes, a recovery helicopter arrived, and Shepard was lifted from the capsule and flown to the aircraft carrier USS Lake Champlain. The entire recovery process took just 11 minutes.

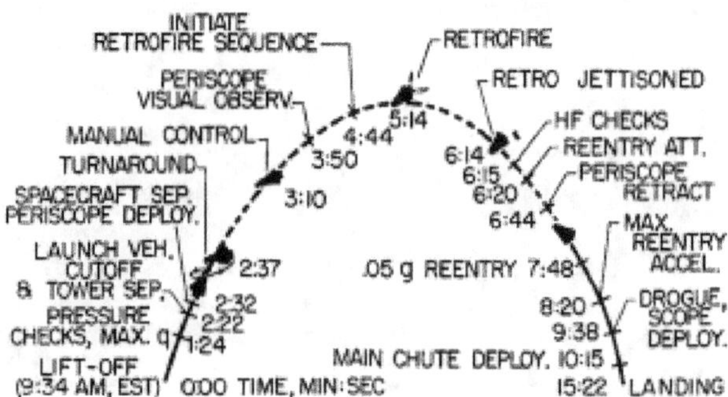

Astronaut Alan B. Shepard Jr. is rescued by a U.S. Marine helicopter from the USS Lake Champlain at the termination of his suborbital flight May 5, 1961, down range from the Florida eastern coast.

Shepard's 15-minute, 22-second flight covered 302 miles (486 km) and reached a speed of 5,180 mph (8,340 km/h). The mission successfully demonstrated the feasibility of human spaceflight, the ability of an astronaut to manually control a spacecraft, and the effectiveness of the Mercury-Redstone launch vehicle. Upon inspection, engineers determined that the Freedom 7 spacecraft was in excellent condition and could have been reused for another mission. Though overshadowed by Gagarin's orbital flight, Shepard's triumph signaled the U.S. commitment to space exploration and set the stage for future missions that would eventually lead to the historic Apollo lunar landings.

Vostok 2

First crewed mission lasting a full day (Gherman Titov).

On August 6, 1961, the Soviet Union launched Vostok 2, marking a significant milestone in the early Space Race. Cosmonaut Gherman Titov became the second human to orbit the Earth, following Yuri Gagarin's historic single-orbit flight aboard Vostok 1 in April of the same year. Unlike Gagarin, who spent less than two hours in space, Titov's mission lasted an entire day. This extended period in space was designed to study the effects of prolonged weightlessness on the human body, a critical question for the future of space exploration.

Russian Kosmonaut Gherman Titov

Sergei Korolev, the chief designer of the Soviet space program, played a significant role in the planning of Vostok 2. Shortly after Gagarin's flight, Korolev began laying out the mission parameters, despite opposition from flight doctors concerned about the effects of long-duration space travel. These concerns were rooted in the results of a previous mission, Korabl-Sputnik 2, which had carried two dogs into space for six orbits. The animals suffered convulsions during the mission, prompting caution for human spaceflights. The medical team proposed a limit of three orbits, but Korolev, convinced that future space missions would require extended time in orbit, argued for a full 24-hour mission.

Practical considerations for the mission's duration also existed. Reentry after three orbits would have placed the spacecraft over Soviet territory, ensuring a safe and controlled landing. However, orbits four through thirteen would bring the capsule over the Pacific Ocean, making recovery difficult or impossible. After the thirteenth orbit, a landing in the Soviet Union was once again feasible, but in Siberia's frozen, remote regions. The mission needed to last at least 24 hours to guarantee a safe recovery in southern Russia.

On August 6, 1961, Vostok 2 lifted off from Baikonur Cosmodrome at 08:57 AM Moscow time. The rocket performed flawlessly, placing Titov into an elliptical orbit with a perigee of 183 km and an apogee of 244 km. During the flight, Titov completed 17.5 orbits of the Earth, surpassing the achievements of both Gagarin and the suborbital flights of American astronauts Alan Shepard and Gus Grissom.

Model of Vostok spacecraft

Although the mission was largely a success, it was not without challenges. Titov suffered from space sickness, becoming the first person to experience what would later be known as space adaptation syndrome. Nausea and vomiting plagued him during the flight, particularly when he tried to eat, complicating his planned experiments. Despite this, Titov took manual control of the spacecraft briefly, marking the first time a cosmonaut had done so. He also captured some of Earth's earliest photographs and motion pictures from space using a Konvas-Avtomat movie camera, providing an invaluable glimpse of the planet from orbit.

Communication with the ground was a key feature of the mission. As he passed over the Soviet Union during his first orbit, Titov exchanged greetings with Soviet Premier Nikita Khrushchev, a repeat of the public relations success from Gagarin's flight. Throughout the mission, telemetry and voice signals from Vostok 2 were picked up by global monitoring stations, allaying suspicions from some quarters that the Soviet spaceflights were being faked.

Reentry, however, proved to be a tense moment. As on Vostok 1, the service module failed to properly detach from the reentry module, causing the combined spacecraft to tumble uncontrollably as it reentered Earth's atmosphere. Aerodynamic forces eventually severed the straps connecting the two sections, allowing Titov to complete reentry safely. As was standard for Vostok missions, Titov ejected from the capsule before landing and parachuted to the ground separately, touching down near the town of Krasny Kut in Saratov Oblast on August 7, 1961, at 07:18 UTC.

The Vostok 2 mission was a triumph for the Soviet space program, solidifying its early lead in the Space Race. Titov, who was just 25 years old then, remains the youngest person to orbit the Earth. His mission demonstrated that humans could survive and function during an extended stay in space, setting the stage for future long-duration missions and advancing the Soviets' space exploration capabilities.

The legacy of Vostok 2 was not without its losses. After the mission, the capsule was reused in 1964 as a ballast weight for a test of the Voskhod capsule's experimental parachute system. The test failed, and Vostok 2 was destroyed, ending its place as a physical artifact of space history. Yet, its impact remained significant, shaping both Soviet space policy and future international space missions.

Two monuments now stand at the landing site near Krasny Kut, commemorating Titov's historic flight. One is a tall, bird-wing-shaped sculpture that points skyward, symbolizing humanity's ascent to the stars. The other is a stone block bearing Titov's likeness, forever linking his name with the early triumphs of space exploration.

Vostok 3 / Vostok 4

First dual crewed spaceflight (Andriyan Nikolayev and Pavel Popovich)

First spacecraft-to-spacecraft radio contact
First simultaneous flight of crewed spacecraft.
First person to float freely in microgravity.

In August 1962, the Soviet Union took another major leap in the Space Race with the successful launches of Vostok 3 and Vostok 4, two crewed spacecraft that orbited the Earth simultaneously. These missions were designed to test human endurance in space and the Soviet ability to manage concurrent spaceflights. The two cosmonauts chosen for this groundbreaking mission were Andriyan Nikolayev, who piloted Vostok 3, and Pavel Popovich, the pilot of Vostok 4.

On August 11, 1962, Vostok 3 launched from the Baikonur Cosmodrome, marking Nikolayev's first spaceflight. He orbited the Earth 64 times over nearly four days, becoming the first person to float freely in space, untethered from his spacecraft seat. This extended stay in space, lasting from August 11 to August 15, set a new record for human spaceflight endurance, one that the United States would not surpass until the Gemini program in 1965.

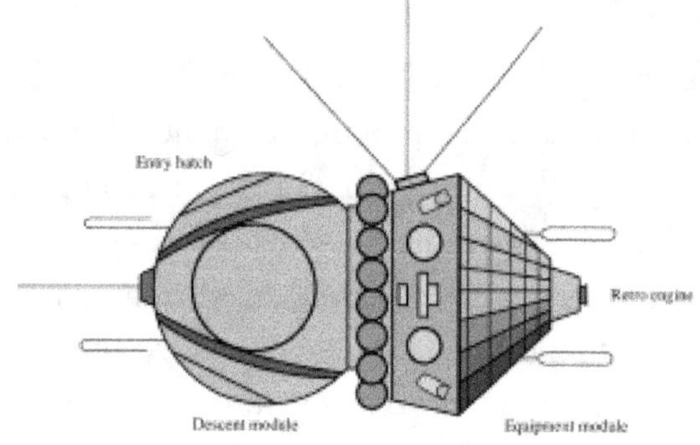

The following day, on August 12, 1962, Vostok 4 was launched with Popovich aboard. He completed 48 orbits of the Earth. Both spacecraft were launched into similar orbits, at altitudes within 3 to 4 kilometers of each other. At one point, the capsules came within visual range of each other, a remarkable achievement at the time. Popovich reported seeing Vostok 3 from orbit, describing it as a "small moon in the distance." This marked the first time that two crewed spacecraft had been in orbit together, a significant milestone in space exploration. Additionally, the cosmonauts communicated via radio, marking the first ship-to-ship communication in space.

The missions were not without their challenges. While the orbits were closely aligned, the Vostok spacecraft did not have the capability for in-orbit maneuvering, which limited their ability to perform an actual rendezvous. As the missions progressed, the distance between the two spacecraft gradually increased, reaching 850 kilometers by the 33rd orbit of Vostok 3 and 2,850 kilometers by its final orbit. Despite this, the Soviet Union gained valuable experience managing two simultaneous crewed spaceflights, a crucial capability for future space missions.

Both Vostok flights included a series of physiological and technical tests aimed at understanding how the human body functioned in space for extended periods. The cosmonauts' physical and mental states were closely monitored through biometric sensors, cabin-mounted cameras, and constant communication with ground control. They also spent time out of their seats conducting tests in weightlessness, which lasted about an hour each day. This data would be essential for planning longer-duration space missions in the future.

While Vostok 4 proceeded largely as planned, there was a malfunction in the life-support system that caused the cabin temperature to drop to a chilly 10°C (50°F). Ground control, misinterpreting a message from Popovich, believed he had requested an early termination of the mission, leading to the spacecraft's premature return. Despite this, the mission was hailed as a success.

Both spacecraft safely re-entered Earth's atmosphere on August 15, 1962, landing just seven minutes apart and approximately 200 kilometers south of Karaganda, Kazakhstan. Nikolayev admitted to reporters that, like his predecessor Gherman Titov on Vostok 2, he had ejected from his capsule and parachuted to Earth separately from the spacecraft. This was in line with Soviet re-entry procedures at the time, though not widely publicized.

The significance of Vostok 3 and Vostok 4 extended beyond the immediate success of placing two cosmonauts in orbit. These missions also provided crucial data about human endurance in space, laying the groundwork for more ambitious Soviet plans. The mission also demonstrated the Soviet Union's capability to perform increasingly complex space operations, as they sought to outpace the United States in the ongoing Space Race.

The decision to pursue dual-mission spaceflights stemmed from lessons learned during previous missions. Gherman Titov's Vostok 2 flight in August 1961, for instance, had exposed cosmonauts to the risks of space sickness, an unknown phenomenon at the time. This spurred Soviet scientists to further investigate the effects of space on the human body. Soviet rocket engineer Sergei Korolev, the mastermind behind the Soviet space program, advocated for a longer spaceflight to explore these effects. Although there were concerns about the health risks, Soviet Premier Nikita Khrushchev gave his approval, allowing the missions to proceed.

The launches of Vostok 3 and Vostok 4 had originally been planned for November 1961, but were delayed due to competing demands on the launch facilities at Baikonur and technical issues with the R-7 rocket. Additional delays occurred in 1962, including concerns over radiation from the U.S. high-altitude nuclear test, Starfish Prime, which temporarily increased the radiation levels in Earth's orbit. By mid-August, conditions were deemed safe enough for launch, and the missions finally took place.

With the successful conclusion of the Vostok 3 and Vostok 4 missions, the Soviet Union once again demonstrated its dominance in the Space Race, having launched more men into space than the United States at that point, and establishing its lead in crewed spaceflight operations. These missions provided invaluable insights into human space endurance, spacecraft operations, and mission management, paving the way for the Soviet Union's continued advancements in space exploration.

Mariner 2

First successful planetary flyby mission (Venus)

Mariner 2, part of the Mariner-Venus 1962 mission, was the first successful American space probe to encounter and transmit data from a planetary body. As the first triumph of NASA's Mariner program, it marked a significant moment in space exploration. The spacecraft was a simplified version of the Block I spacecraft used in the Ranger program, designed to be an exact copy of Mariner 1, whose mission had failed. Together, these missions are often called the Mariner R missions.

Initially, NASA planned to launch Mariner 1 and Mariner 2 using the Atlas-Centaur rocket. However, technical difficulties with the Centaur stage forced NASA to switch to the smaller Atlas-Agena B, a rocket with significantly less lift capacity. As a result, the design of the Mariner spacecraft had to be greatly simplified, carrying far less instrumentation compared to the Soviet Venera probes of the same period. For instance, Mariner 2 was not equipped with a television camera, a trade-off required by the limitations of the Atlas-Agena B's reduced capacity compared to the more powerful Soviet 8K78 booster.

On August 27, 1962, Mariner 2 was launched from Cape Canaveral, Florida, and embarked on its historic journey toward Venus. On December 14, 1962, it passed within 34,773 kilometers (21,607 miles) of the planet, successfully collecting and transmitting valuable scientific data back to Earth.

The spacecraft itself featured a hexagonal bus with a diameter of 100 cm (39.4 in), from which solar panels, instrument booms, and antennas were extended. The suite of scientific instruments aboard Mariner 2 included two radiometers (operating in the microwave and infrared spectrums), a micrometeorite sensor, a solar plasma sensor, a charged particle detector, and a magnetometer. These instruments were designed to study Venus' atmospheric properties, surface temperature distribution, as well as the broader environment of interplanetary space.

Mariner 2's primary mission objectives were to transmit communications while in the vicinity of Venus and to perform radiometric temperature measurements of the planet. A secondary objective was to investigate the interplanetary magnetic field and the charged particle environment between Earth and Venus. During its journey, Mariner 2 made groundbreaking discoveries. It confirmed the existence of the solar wind—a constant stream of charged particles flowing outward from the Sun—previously observed by the Soviet Luna 1 in 1959. The probe also found that interplanetary dust was far less abundant than expected and detected bursts of high-energy particles from solar flares as well as cosmic rays originating from outside the Solar System.

As Mariner 2 flew past Venus, it used its radiometers to scan the planet, revealing that Venus had sweltering surface temperatures and cooler cloud layers. This data significantly advanced the understanding of Venus, confirming that its surface was far hotter than previously thought, likely due to a runaway greenhouse effect.

Mariner 2

Mariner 2's mission occurred during the intense geopolitical rivalry between the United States and the Soviet Union, which drove the Space Race. Both nations viewed space exploration as a demonstration of technological, military, and political superiority. In 1957, the Soviet Union launched Sputnik 1, the first artificial satellite to orbit Earth, triggering a wave of American efforts to catch up. The United States successfully launched its own satellite,

Explorer 1, in February 1958, following the Soviet launch of Sputnik 2, which carried the first living creature into space—Laika the dog.

Once Earth orbit had been conquered, attention quickly turned to the Moon. The Pioneer program, America's early attempt to reach the Moon, faced several setbacks, with three unsuccessful lunar missions in 1958. Meanwhile, the Soviet Union achieved a major milestone in 1959 with the Luna 1 probe, which became the first spacecraft to fly past the Moon, followed by Luna 2, which impacted the lunar surface, becoming the first human-made object to reach the Moon.

With lunar exploration progressing, both superpowers set their sights on Venus, the closest planet to Earth. The timing of Venus' orbits relative to Earth creates launch windows every 19 months that allow for optimal fuel efficiency in sending spacecraft between the two planets. The first such opportunity in 1957 passed before either nation had the necessary technology. By June 1959, the second window approached, and U.S. contractor Space Technology Laboratory (STL) planned to take advantage of it. However, the technology wasn't ready in time, and the opportunity was missed. The probe designed for this mission, Pioneer 5, was repurposed and launched in March 1960, successfully transmitting data from deep space but not making a planetary flyby.

The Soviets, however, made their first attempt to reach Venus with the Venera 1 probe, launched on February 12, 1961. Although Venera 1 became the first spacecraft to fly by Venus, it stopped transmitting before reaching the planet, leaving the opportunity for further planetary exploration open.

In July 1960, NASA contracted the Jet Propulsion Laboratory (JPL) to develop a spacecraft capable of reaching Venus during the next launch window in 1962. Initially called "Mariner A," the spacecraft was to be launched using the Atlas-Centaur rocket. By mid-1961, it became evident that the Centaur would not be ready for the mission. JPL proposed a lighter spacecraft version that could be launched on the Atlas-Agena B. This revised design, known as "Mariner R," was a hybrid of the Mariner A and JPL's Ranger lunar explorer. NASA approved the plan, and JPL embarked on an 11-month crash program to ready the spacecraft.

The Mariner R program, designed to explore Venus, saw the construction of three spacecraft: two for launch and one for testing, which also served as a spare. These spacecraft were tasked with transmitting data from distances exceeding 26 million miles (42 million kilometers) while withstanding solar radiation twice as intense as in Earth's orbit. This monumental challenge required precise engineering to ensure their success in the harsh environment of interplanetary space.

Each Mariner R spacecraft, including Mariner 2, weighed around 447 pounds (203 kilograms), of which only 41 pounds (18 kilograms) were dedicated to scientific instruments. The remaining mass comprised essential systems such as propulsion, maneuvering fuel, and communication equipment. Once fully deployed in space, with its solar panels extended, Mariner measured 12 feet (3.7 meters) in height and spanned 16.5 feet (5 meters) across.

The spacecraft featured a hexagonal central body, housing six distinct cases of electronic and mechanical equipment:

Two cases housed the power system, which included switchgear to manage the energy generated by 9,800 solar cells. These cells charged a 33-pound (15.1 kg), 1,000-watt silver-zinc storage battery that powered the spacecraft.

Two additional cases held the radio receiver, the 3-watt transmitter, and control systems for the scientific instruments.

A fifth case was dedicated to digitizing the analog data received from the instruments, preparing it for transmission to Earth.

The final case housed three gyroscopes, which helped maintain the spacecraft's orientation. It also contained the central computer, known as the spacecraft's "brain," which controlled all activities based on pre-programmed instructions.

A monopropellant hydrazine rocket motor propelled course corrections at the spacecraft's rear. Additionally, a nitrogen gas stabilizing system, controlled by onboard gyroscopes and sensors for the Sun and Earth, ensured that Mariner maintained the correct orientation to send and receive data from Earth. The spacecraft's high-gain parabolic antenna was mounted on the underside and aimed at Earth to ensure a strong communication signal. An omnidirectional antenna served as a backup, transmitting when the spacecraft lost its orientation, albeit with a weaker signal.

Temperature control was a critical aspect of Mariner's design. It relied on both passive methods, such as insulating and highly reflective surfaces, and active methods, including louvers that protected the onboard computer from overheating. However, since there were no test chambers capable of simulating the intense solar environment near Venus, the effectiveness of these cooling methods could only be fully tested during the actual mission.

When the Mariner project began, relatively little was known about Venus due to its dense and opaque atmosphere, which blocked telescopic observations of the planet's surface. Scientists were uncertain about key characteristics, such as whether Venus had water beneath its clouds or if the planet was tidally locked with the Sun, as Earth's Moon is with Earth. Although radar observations suggested Venus rotated very slowly, the planet's precise rotation rate remained unclear. Additionally, scientists had detected significant levels of carbon dioxide in Venus' atmosphere—500 times greater than Earth's—which hinted at a possible runaway greenhouse effect with surface temperatures as high as 600 K (327 °C; 620 °F).

Mariner's mission was designed to answer these questions. The spacecraft was equipped with instruments to measure the planet's surface temperature and atmospheric properties, as well as to search for a potential magnetic field and analogs to Earth's Van Allen radiation belts. Additionally, as Mariner spent most of its journey to Venus in interplanetary space, it offered an opportunity to study the solar wind and the distribution of cosmic dust, contributing valuable data about space conditions beyond Earth.

The primary scientific instruments aboard Mariner 2 included:

Microwave Radiometer: This two-channel instrument operated in two frequency bands (13.5 mm and 19 mm) and was designed to measure Venus' surface temperature and atmospheric properties. It provided crucial data on the temperature differences between the planet's day and night sides and the region along the terminator—the line dividing day from night.

Infrared Radiometer: Measuring in two spectral ranges, this instrument detected the effective temperatures of small areas on Venus, which could include surface radiation, atmospheric conditions, or cloud layers. The infrared radiometer was particularly focused on identifying temperatures between 200 K and 500 K, helping to verify the greenhouse effect hypothesis.

Magnetometer: This instrument measured magnetic fields in both interplanetary space and near Venus, offering insight into the magnetic environment around the planet.

Ionization Chamber and Geiger-Müller Tubes: These devices were used to detect high-energy cosmic radiation and charged particles, particularly near Venus.

Cosmic Dust Detector: This instrument measured the density of cosmic dust particles in space, providing valuable data on the distribution of these particles beyond Earth's orbit.

Solar Plasma Spectrometer: This instrument studied the spectrum of charged particles in the solar wind, helping scientists understand the behavior of solar particles as they traveled through space.

In addition to these instruments, Mariner 2 included systems for managing and transmitting the data collected. A data conditioning system (DCS) gathered and converted the scientific data, while a scientific power switching (SPS) unit controlled the power supplied to the radiometers and other instruments, ensuring that they operated efficiently.

Interestingly, Mariner 2 did not carry a camera. Given the spacecraft's weight restrictions, project scientists deemed a camera unnecessary, believing Venus' thick cloud cover would prevent useful images from being captured.

Carl Sagan, an influential planetary scientist involved with the mission, argued for including a camera, suggesting that gaps in Venus' cloud layer might allow for visual surface observation. Despite his advocacy, the camera was excluded, and Mariner 2 focused solely on its scientific objectives.

The launch window for the Mariner 2 mission, determined by the relative orbits of Earth and Venus and constrained by the capabilities of the Atlas-Agena rocket, spanned 51 days from July 22 to September 10, 1962. This timeframe allowed NASA to launch two operational spacecraft within 30 days, each following slightly different trajectories toward Venus. Both spacecraft were scheduled to arrive at the planet between December 8 and 16, 1962. Given that Cape Canaveral's Launch Complex 12 was the only available site for Atlas-Agena launches, it required 24 days to prepare for a second launch, leaving a tight 27-day margin for error.

Each Mariner spacecraft was launched into a temporary parking orbit, after which the Agena second stage would ignite a second time to send the probe on its interplanetary trajectory. Errors in the trajectory were corrected mid-course using Mariner's onboard propulsion system. During these crucial phases, stations at Ascension Island and Pretoria, South Africa provided real-time radar tracking with additional optical tracking from Palomar Observatory. Deep space communications relied on three tracking stations in Goldstone, California, Woomera, Australia, and Johannesburg, South Africa. These stations, spaced 120 degrees apart around the globe, ensured continuous coverage of the spacecraft's journey to Venus.

Mariner 1, the first spacecraft launched as part of the mission, met a disastrous fate. On July 22, 1962, its Atlas-Agena rocket veered off course due to a combination of a defective signal from the Atlas and a coding error in the ground-based guidance computer. The Range Safety Officer was forced to destroy the spacecraft shortly after launch.

Just two days after the failure of Mariner 1, Mariner 2 and its Atlas-Agena rocket were rolled out to Launch Complex 12. Preparation for launch proved challenging, with the Atlas rocket's autopilot suffering multiple malfunctions, including severe damage to the servo amplifier due to short-circuited transistors. Despite these hurdles, Mariner 2 was successfully launched on August 27, 1962, at 1:53 AM EST (06:53:14 UTC). Interestingly, the bug that doomed Mariner 1 was still present in the rocket's software at the time of Mariner 2's launch. However, since the faulty code only activated when the ground data feed was interrupted—and no such interruptions occurred—Mariner 2 launched without issue.

The launch proceeded smoothly until the Agena booster engine cut off, causing the V-2 vernier engine to lose pitch and yaw control, leading to the vehicle's rapid roll. This instability nearly compromised the spacecraft's structural integrity. After 189 seconds of roll, the malfunctioning feedback transducer, which had caused the issue, was pushed back into place by centrifugal forces. Miraculously, the Atlas rocket regained stability, and the launch continued as planned.

At T+26 minutes, 3 seconds, Mariner 2 separated from the Agena stage and was injected into a geocentric escape trajectory. The NASA tracking station in Johannesburg, South Africa, acquired Mariner 2's signal 31 minutes after liftoff. Solar panels deployed 44 minutes post-launch, and the spacecraft locked onto the Sun, ensuring stable power generation.

On August 29, just two days after launch, Mariner 2 began activating its cruise science experiments. By September 3, the spacecraft initiated its Earth acquisition sequence, establishing a stable lock on Earth for communication.

Although Mariner 2's initial trajectory was slightly off due to the Atlas-Agena launch, a mid-course correction was scheduled for September 4, 1962. The spacecraft executed a roll-turn, pitch-turn, and motor-burn sequence that took about 34 minutes. This maneuver successfully corrected its course, although the spacecraft temporarily lost its Sun and Earth locks. Both were reacquired within a few hours, ensuring Mariner 2 remained on track.

However, Mariner 2 experienced several attitude control issues during its journey. On September 8, the spacecraft's attitude control system automatically activated its gyros, possibly due to a sensor malfunction or a collision with a small object, causing the cruise science experiments to shut down temporarily. Another similar issue occurred on September 29, but again the spacecraft recovered, with the Earth sensor returning to normal functionality.

Solar Panel Failures and Power Management

On October 31, 1962, one of Mariner 2's solar panels suddenly suffered a partial short circuit, leading mission controllers to shut down the cruise science instruments as a precaution. Though the panel briefly resumed normal operation a week later, it permanently failed on November 15. Fortunately, Mariner 2 was close enough to the Sun that its remaining panel could generate sufficient power for the spacecraft to continue its mission.

After 110 days in flight, Mariner 2 became the first spacecraft to successfully encounter another planet, passing within 34,773 kilometers (21,607 miles) of Venus on December 14, 1962. The spacecraft's instruments operated flawlessly during the flyby, collecting and transmitting valuable data back to Earth.

Following its historic flyby, Mariner 2 continued its journey, entering cruise mode once again. The spacecraft passed its closest point to the Sun, or perihelion, on December 27, 1962, at 105,464,560 kilometers (65,532,640 miles). The final transmission from Mariner 2 was received on January 3, 1963, bringing the total duration of the mission to 129 days. After completing its mission, Mariner 2 entered heliocentric orbit.

Mariner 2's scientific discoveries were groundbreaking. The spacecraft confirmed the extremely high temperatures of Venus' atmosphere, which were measured to be around 500°C (773 K; 932°F). It also provided the first in-situ measurements of the solar wind, a constant stream of charged particles flowing from the Sun, confirming the predictions made by physicist Eugene Parker. Additionally, Mariner 2 recorded valuable data about cosmic dust density in interplanetary space, which was found to be much lower than near Earth.

As Mariner 2 flew past Venus, its microwave radiometer performed three scans of the planet. The first scan covered the dark side, the second near the terminator, and the third the light side. The measurements revealed that the temperature across Venus remained fairly consistent, with peak values of 490 K (217°C; 422°F) on the dark side and 511 K (238°C; 457°F) on the light side. These results supported the hypothesis of an optically thick atmosphere and a surface temperature driven by a runaway greenhouse effect.

The infrared radiometer confirmed earlier Earth-based observations of Venus' temperature, with no significant difference between the light and dark sides. This suggested that heat was efficiently transported across the planet, likely due to strong atmospheric circulation.

Mariner 2's magnetometer detected a persistent interplanetary magnetic field ranging from 2 to 10 nanotesla, aligning with previous Pioneer 5 observations. However, no magnetic field was detected near Venus, indicating that if the planet did have a magnetic field, it was at least 10 times weaker than Earth's. This finding was later confirmed by the Pioneer 12 mission in 1980, which detected only a faint magnetic field around Venus.

The success of Mariner 2 was a monumental achievement in space exploration, not only marking the first successful planetary flyby but also providing invaluable data about Venus, the solar wind, and interplanetary space. Mariner 2 set the stage for future planetary missions, solidifying NASA's role in the Space Race and advancing humanity's understanding of the solar system.

Vostok 6

First woman in space (Valentina Tereshkova)
First civilian in space

On June 16, 1963, Vostok 6, a significant mission in the history of space exploration, became the first human spaceflight to carry a woman into space. Valentina Tereshkova, a Soviet cosmonaut, made history as she orbited the Earth, opening new frontiers for women in science and space exploration. This mission not only marked a significant achievement in the Soviet space program but also contributed valuable data on the effects of space travel on the female body.

The launch of Vostok 6 proceeded smoothly, in contrast to Vostok 5, which had been delayed due to technical issues. Tereshkova's spacecraft orbited the Earth for nearly three days, during which she conducted various tasks similar to those of her predecessors. She maintained a flight log, captured photographs of the Earth's horizon, and manually oriented the spacecraft. Notably, her photographs provided insights into atmospheric aerosol layers, an important discovery that contributed to the understanding of Earth's atmosphere.

Initially, the Soviet space agency had planned for a joint mission in which two Vostok spacecraft, each carrying a female cosmonaut, would be launched. However, the mission was scaled back due to program cutbacks and the impending transition to the Voskhod program. Instead, Vostok 6 flew alongside Vostok 5, which was piloted by Valery Bykovsky, marking the final flight of the Vostok program and the last mission of the Vostok 3KA spacecraft series.

Tereshkova's return to Earth was just as eventful as her time in orbit. Like all Vostok cosmonauts, she had to eject from the capsule and descend by parachute. She landed safely near Baevo, Russia, approximately 200 kilometers west of Barnaul. A statue now stands at the landing site, commemorating her historic achievement.

Post-flight reports revealed some of the challenges Tereshkova faced during her mission. While the Soviet state television network broadcast live footage of her from inside the spacecraft, official accounts of her condition were later described as evasive. Tereshkova reported experiencing body pains, discomfort with her helmet headset, and nausea caused by the taste of space food. Additionally, a significant oversight by the mission planners resulted in her being provided with food, water, and toothpaste—but no toothbrush. Despite these issues, Tereshkova completed her mission successfully, though official Soviet accounts rated her performance as only "adequate" rather than "outstanding."

A more serious issue arose when Tereshkova discovered a programming error in the spacecraft's control system that caused the ship to ascend instead of descend. She reported the problem to the spacecraft's chief designer, Sergey Korolev, who instructed her to manually correct the trajectory by entering new data into the descent program. The details of this error remained a secret until 2004, when Tereshkova publicly revealed it.

The Vostok 6 mission was a groundbreaking achievement, not just for the Soviet space program, but for the broader narrative of human space exploration. Valentina Tereshkova's bravery and perseverance paved the way for future generations of women in space. The capsule she flew in now resides at the RKK Energia Museum in Korolyov, near Moscow, and was also featured in the "Cosmonauts" exhibition at the Science Museum in London, serving as a testament to the daring spirit of the Space Race.

Mars 1

First Mars flyby, although contact was lost

On November 1, 1962, the Soviet Union launched Mars 1, an ambitious automatic interplanetary station aimed at conducting a flyby of Mars, coming within approximately 11,000 kilometers (6,800 miles) of the planet's surface. This mission, also known as 1962 Beta Nu 1, Mars 2MV-4, and Sputnik 23, represented a pioneering effort in the Soviet Mars probe program. It was designed to capture and transmit groundbreaking data, including detailed images of the Martian surface, as well as information about cosmic radiation, micrometeoroid impacts, Mars' magnetic field, atmospheric composition, radiation environment, and the potential for organic compounds on the planet.

Following a successful launch, Mars 1 separated from its booster's fourth stage and deployed its solar panels. However, early telemetry indicated a critical issue: a leak in one of the gas valves in the spacecraft's orientation system. As a result, the spacecraft was transitioned to gyroscopic stabilization to maintain its trajectory toward Mars.

Over the next several months, Mars 1 sent valuable interplanetary data back to Earth, completing sixty-one radio transmissions. These transmissions were initially held at two-day intervals, later extended to five-day gaps, collecting data that contributed to the understanding of space beyond Earth's orbit.

Unfortunately, communication was lost on March 21, 1963, when the spacecraft was approximately 106,760,000 kilometers (66,340,000 miles) from Earth and en route to Mars. The most likely cause was the failure of the spacecraft's orientation system, which rendered it unable to point its antenna toward Earth for continued transmissions. Despite this setback, Mars 1 completed its closest approach to Mars on June 19, 1963, passing at a distance of about 193,000 kilometers (120,000 miles) from the planet. After this flyby, the spacecraft entered a heliocentric orbit around the Sun.

The spacecraft, cylindrical in shape, measured 3.3 meters (11 feet) in length and 1 meter (3.3 feet) in diameter, expanding to 4 meters (13 feet) with its solar panels deployed. It was based on the Venera design, a proven model used for Venus exploration missions. Mars 1's structure was divided into two compartments: the upper orbital module housed the guidance and propulsion systems, while the lower experimental module contained the scientific instruments. Power was generated through two solar panels with a total surface area of 2.6 square meters (28 square feet) and stored in a 42 ampere-hour cadmium-nickel battery.

However, Mars 1's journey to the Red Planet was not without its challenges. Shortly after leaving Earth's orbit, the spacecraft encountered a malfunction in its orientation system. A gas valve leak forced mission control to switch the spacecraft to gyroscopic stabilization. Despite this, Mars 1 continued to send valuable data from deep space. Throughout its mission, it made sixty-one radio transmissions, gradually transitioning from two-day intervals to five-day intervals as it moved further from Earth.

Mars 1's scientific mission was cut short on March 21, 1963, when communication was lost as the spacecraft was approximately 106.7 million kilometers (66.3 million miles) from Earth. The failure was likely due to the spacecraft's antenna orientation system malfunction. Though Mars 1 never achieved its intended flyby, it came within 193,000 kilometers (120,000 miles) of the planet on June 19, 1963, before continuing on its unintended orbit around the Sun.

The scientific instruments aboard Mars 1 were state-of-the-art for their time. The spacecraft was equipped with multiple communication systems, operating at wavelengths of 1.6 meters, 32 centimeters, and higher-frequency ranges between 5 and 8 centimeters. A high-gain parabolic antenna was used to transmit data, while backup systems ensured that communication could be maintained through various antennas, including omnidirectional ones located on the solar panels. The spacecraft's imaging system was also designed to capture television-like images of Mars' surface, though these never reached Earth.

Despite losing contact, Mars 1 contributed valuable scientific data during its flight. It measured the density of micrometeoroids in space, particularly noting increased impacts from the Taurids meteor shower at distances between 6,000 and 40,000 kilometers (3,700 and 24,900 miles) from Earth. Magnetic field intensities in interplanetary space were recorded, with peaks as high as 9 nanoteslas (nT). Mars 1 also confirmed the existence of the solar wind and provided critical data on Earth's radiation belts, helping to map their extent and confirming earlier measurements.

Mars 1's mission, while falling short of its goal, represented an important milestone in planetary exploration. It demonstrated the Soviet Union's ability to send spacecraft beyond Earth's orbit and laid the groundwork for future Mars missions. Despite being designated Mars 1, it was preceded by several failed missions, including Mars 2MV-4 No.1,

X-15 Flight 90

First reusable piloted spacecraft
First spaceplane (suborbital)

The North American X-15 program was one of the most ambitious and groundbreaking aviation efforts during the Space Race, contributing significant advancements to both aeronautics and astronautics. Conducted by NASA and the U.S. Air Force, the X-15 program was designed to explore the edges of space and test the limits of flight beyond the Earth's atmosphere. Pilots such as Joseph A. Walker, Robert White, and William Knight pushed the boundaries of what was possible, flying at extreme altitudes and velocities that laid the groundwork for future space exploration.

Flight 35: First Flight into the Mesosphere

The X-15's early flights aimed to gather crucial data about high-speed flight in the mesosphere, a region of the Earth's atmosphere located between 50 and 85 kilometers above sea level. On Flight 35, pilot Joseph A. Walker became the first person to enter the mesosphere with the X-15, marking a significant milestone in high-altitude research. This flight provided essential insights into the effects of high-speed travel at these altitudes, where traditional aircraft systems and materials faced new challenges.

On Flight 62, pilot Robert White flew the X-15 to a record altitude of 95.94 kilometers, achieving what was recognized as the first U.S. spaceflight. While White's mission did not cross the internationally defined Kármán line, it marked a significant achievement in the U.S. space program, demonstrating that manned spaceplanes could travel to the edge of space and return safely.

Piloted by Robert Rushworth, Flight 87 further extended the program's reach into space. Although Rushworth did not surpass the Kármán line, his flight reached an altitude of 85.5 kilometers, continuing the X-15's series of high-altitude spaceflights. This mission provided valuable data on the behavior of the aircraft and the human body at near-space altitudes.

On July 19, 1963, Flight 90, piloted by Joseph A. Walker, made aviation history by becoming the first X-15 flight to pass the 100-kilometer-high Kármán line, the internationally recognized boundary of space. This mission was a remarkable milestone, making Walker the first U.S. civilian to fly into space. Launched from an NB-52B bomber at 17:19 UTC over Smith Dry Lake, Nevada, the X-15 ascended to a maximum altitude of 106.01 kilometers (347,800 feet) and reached a speed of 5,971 km/h (Mach 5.50). During the flight, Walker experienced several minutes of weightlessness before re-entering the Earth's atmosphere, where the aircraft endured temperatures of up to 650°C due to re-entry heating.

The flight included a series of scientific experiments, including deploying an 80 cm diameter balloon to measure air density and testing of horizon scanners and photometers. However, the balloon instrumentation failed, limiting the scientific return of this experiment. The flight lasted 11 minutes and 24 seconds, with the X-15 landing safely at Rogers Dry Lake, Edwards Air Force Base, at 19:04 UTC.

Flight 91: Program Altitude Record

One month after Flight 90, Joseph A. Walker set another program record on Flight 91, reaching an even higher altitude of 107.8 kilometers, surpassing his previous achievement. This marked the highest altitude ever reached by the X-15 program and solidified its role in the Space Race as a precursor to manned orbital spaceflights. Walker's flights demonstrated the feasibility of controlled spaceplane re-entries, an important consideration for future space missions.

In another significant achievement, William J. "Pete" Knight set the program speed record on Flight 188, reaching an astonishing Mach 6.72 (7,273 km/h). This speed remains one of the highest ever achieved by a manned aircraft. The data collected during these high-speed flights were vital for developing future spacecraft, particularly in understanding the thermal and structural stresses experienced at extreme velocities.

The X-15 program was not without its dangers. On November 15, 1967, Flight 191, piloted by Michael J. Adams, ended in tragedy. Adams encountered control issues while flying at high altitude, and despite his efforts to regain control, the X-15 broke apart during re-entry, resulting in his death. This was the only fatality in the X-15 program, highlighting the inherent risks of pushing the boundaries of human flight.

The X-15 program, which ran from 1959 to 1968, made invaluable contributions to both manned spaceflight and aeronautics. Its 199 flights, piloted by astronauts and engineers such as Neil Armstrong, Michael Adams, and Joe Engle, produced extensive research on spaceplane re-entry, high-speed flight, and human endurance in extreme conditions. These efforts advanced scientific knowledge and directly influenced the design and operation of later space programs, including the Space Shuttle.

Contributions of North American Aviation and Key Pilots

North American Aviation played a crucial role in the success of the X-15 program, providing the technology and expertise required to achieve these remarkable feats. The company's engineers and test pilots, including Neil Armstrong, Michael Adams, and Joe Engle, were instrumental in pioneering the research necessary to understand flight at the edge of space. Pilots like Joseph A. Walker and Robert White earned their places in space history through their bravery and skill, conducting missions that pushed both man and machine to their limits.

Syncom

USA
First geosynchronous satellite
First geostationary satellite

The Syncom program, short for "synchronous communication satellite," marked a significant moment in the Space Race by pioneering geosynchronous communication technology. Initiated in 1961 as a NASA program, the Syncom series consisted of experimental satellites developed by the Space and Communications Division of Hughes Aircraft Company, now part of Boeing. The Syncom satellites were the first to demonstrate the feasibility of synchronous and later geostationary orbits for communication purposes, a breakthrough that revolutionized global communication.

Syncom 1, launched on February 14, 1963, was intended to be the world's first geosynchronous communication satellite. However, shortly after the spacecraft's apogee motor fired to place it in its intended orbit, contact was lost due to an electronics failure. Despite its failure, Syncom 1's orbital path was later verified by telescopic observations, confirming that it had achieved an almost 24-hour orbit, albeit with a 33-degree inclination. The lessons learned from this mission would shape the success of subsequent launches.

On July 26, 1963, Syncom 2 was successfully launched aboard a Thor Delta B rocket from Cape Canaveral, becoming the world's first operational geosynchronous communication satellite. Syncom 2 operated at an altitude calculated decades earlier by space pioneer Herman Potočnik Noordung. During its first year, NASA conducted various communication tests, including voice, teletype, and facsimile transmissions. It also facilitated the first two-way live telephone conversation between heads of state via satellite when U.S. President John F. Kennedy spoke to Nigerian Prime Minister Abubakar Tafawa Balewa aboard the USNS Kingsport in Lagos Harbor. This marked a major satellite communication milestone, demonstrating its global connectivity potential.

In September 1963, Syncom 2 relayed the first test television transmission through a geosynchronous satellite. Although the video was of low quality and lacked audio, it represented a key moment in satellite television broadcasting.

Syncom 3, launched on August 19, 1964, took satellite communications further by becoming the world's first geostationary satellite, positioned near the International Date Line. Equipped with a wideband channel for television broadcasting, Syncom 3 transmitted live coverage of the 1964 Summer Olympics in Tokyo to the United States, an unprecedented event in global broadcasting. Though Syncom 3 is sometimes credited with being the first satellite to transmit television signals across the Pacific, that distinction technically belongs to Relay 1, which broadcast from the U.S. to Japan in 1963.

By the end of 1964, NASA had completed its research and development work with the Syncom satellites. On January 1, 1965, operational control of Syncom 2 and 3 was transferred to the U.S. Department of Defense (DOD). The satellites became integral to military communications, particularly during the Vietnam War. Although Syncom 3 was deactivated in 1969, it remains in geosynchronous orbit, having drifted to 123 degrees west longitude over five decades.

In the 1980s, the Syncom series was revived and evolved into the much larger Syncom IV satellites under the Leasat (Leased Satellite) program. Hughes Aircraft Company designed and manufactured these satellites leased to the U.S. military for secure communication purposes. The Leasat satellites, also known as HS 381, were significantly larger than their predecessors, weighing over 7 tons with fuel and measuring 14 feet in height. They were the first satellites explicitly designed for deployment from the Space Shuttle payload bay, launched by the shuttle in a distinctive "Frisbee" style.

The five satellites in the Leasat series—designated F1 through F5—provided vital communication services for the U.S. Navy, Marine Corps, Air Force, and Army. Each satellite featured a wideband UHF channel, multiple relay channels, and broadcast capabilities in the military's X-band. Though primarily used by the U.S. military, the Australian Defence Force later leased services from the final operational Leasat satellite.

Not all Leasat launches were successful. In 1985, Leasat F3 failed to initiate its transfer to geostationary orbit after its release from the Space Shuttle Discovery. Despite efforts by astronauts to manually repair the satellite during a spacewalk, it was left stranded in low Earth orbit. However, later that year, NASA executed the first space salvage mission. During Shuttle mission STS-51-I, astronauts captured Leasat F3, manually spun it down, and installed a bypass device, enabling it to reach geostationary orbit. This mission set a precedent for future satellite repairs, including the 1992 recovery of Intelsat 603.

Leasat F4 also encountered issues shortly after its deployment, but lessons from F4's failure helped prolong the operational life of other Leasat satellites. In fact, Leasat F4 was reactivated years later to test alternative propellants, showcasing Hughes engineers' innovative use of non-traditional fuels for station-keeping maneuvers.

Leasat F5, the last satellite in the series, was launched in January 1990 aboard Space Shuttle Columbia during mission STS-32. Serving the U.S. and Australian military, it remained operational for an impressive 25 years before decommissioning in 2015. The Leasat program represented a significant leap forward in military satellite communications, laying the groundwork for future generations of secure satellite networks.

Voskhod 1

First spaceflight to carry more than one crewman into orbit (3)

Voskhod 1: The Dawn of Multi-Crewed Spaceflight

On October 12, 1964, the Soviet Union made history once again with the launch of Voskhod 1, the seventh crewed Soviet space flight. This mission marked several milestones in the annals of human space exploration. Flown by cosmonauts Vladimir Komarov, Konstantin Feoktistov, and Boris Yegorov, Voskhod 1 became the first spaceflight to carry more than one human into orbit. It was also the first mission to fly without spacesuits, making the voyage both groundbreaking and extraordinarily risky. The spacecraft reached a record altitude of 336 kilometers (209 miles) before safely returning to Earth on October 13, 1964.

The launch of Voskhod 1 was a significant moment in the ongoing Space Race, coming at a time when both the Soviet Union and the United States were vying for supremacy in space exploration. This mission underscored the Soviet Union's ambition to push the boundaries of human spaceflight, even under conditions that would later be regarded as precarious.

The original design of the Voskhod spacecraft was based on the earlier Vostok capsule, which had been engineered to accommodate only one cosmonaut. However, in response to political pressure and the desire to outpace American achievements in space, Soviet engineers were tasked with modifying the capsule to fit three crew members. This necessitated the removal of the ejection seats used in previous missions and replacing them with three closely arranged couches, leaving no room for spacesuits. The cosmonauts were forced to go on strict diets to fit into the cramped confines of the capsule.

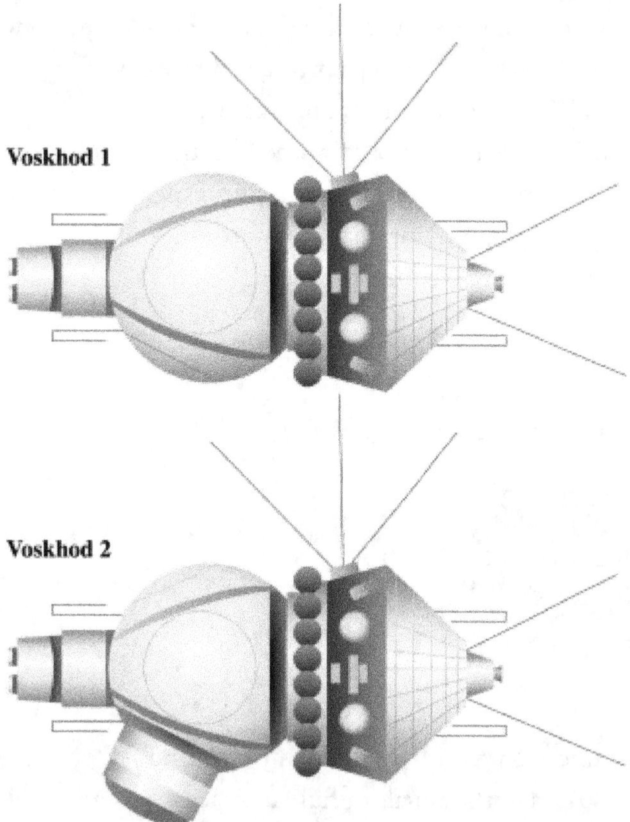

Despite these limitations, Voskhod 1 carried a symbolic payload—a ribbon from the Communard banner of the 1871 Paris Commune—into orbit, underscoring the political overtones that accompanied many Soviet space missions. The cosmonauts on board included Vladimir Komarov as the command pilot, making his first spaceflight; Konstantin Feoktistov, the first engineer to fly into space; and Boris Yegorov, a medical doctor, and the first physician in space. Yegorov's inclusion in the mission was seen as a significant step in advancing biomedical research in space.

The selection of the Voskhod 1 crew was not without controversy. The original crew, which included Boris Volynov, Georgi Katys, and Boris Yegorov, was abruptly replaced just days before the launch. Katys was removed after the KGB discovered that his father had been executed during the Great Purge of 1937, and Volynov, who was of part-Jewish heritage, faced discrimination. Although Volynov would later go on to fly on Soyuz 5 and Soyuz 21, his exclusion from Voskhod 1 highlighted the influence of politics in Soviet space missions.

Sergei Korolev, the chief designer of the Soviet space program, strongly advocated for having engineers aboard spacecraft, believing that those who designed the vessels should also experience spaceflight. This belief led to the selection of Konstantin Feoktistov, who had contributed to the design of the Vostok, Voskhod, and Soyuz spacecraft. Political connections also played a role in Yegorov's selection, as his father's influence in the Soviet Politburo helped secure his place on the mission.

The Voskhod spacecraft itself was a heavily modified version of the earlier Vostok capsule. The ejection seat was removed, and a backup solid-fuel retrorocket was added to the descent module, along with a new braking system to soften the landing. Due to space and weight limitations, the crew did not wear spacesuits, a decision that added to the risks of the mission. The cosmonauts would not have survived if the capsule had depressurized in orbit.

Liftoff occurred at 7:30 AM Moscow time on October 12, 1964. Much of the 24-hour mission was devoted to biomedical research and studying how a multidisciplinary team—composed of a pilot, an engineer, and a physician—could work together in space. The mission provided valuable insights into the physiological and psychological effects of space travel on humans, especially in such confined quarters.

However, the mission was not without its political intrigue. While the cosmonauts were orbiting Earth, Soviet Premier Nikita Khrushchev was removed from power. Just hours after speaking with the cosmonauts via radio, Khrushchev was summoned back to Moscow, where he learned of his ouster from office. When the crew returned to Earth, they were greeted by Leonid Brezhnev and Alexei Kosygin, who had just assumed leadership of the Soviet Union.

Despite the mission's success, Soviet space officials privately referred to Voskhod 1 as a "circus." The rushed crew selection process, the dangerous lack of spacesuits, and the political upheaval during the flight all contributed to a sense of chaos surrounding the mission. Nevertheless, the mission significantly, albeit temporarily, impacted the global perception of the Soviet space program. NASA Administrator James E. Webb called the flight a "significant space accomplishment," acknowledging that it demonstrated the Soviet Union's commitment to maintaining its leadership in space exploration.

The development of Voskhod 1 began in April 1964, and the mission required extensive modifications to the Vostok spacecraft and the 11A57 launch vehicle. The most significant changes included the removal of the ejection seat and the addition of three crew couches with suspension systems to absorb the shock of launch and landing. To compensate for the additional weight, the launch vehicle's third stage was upgraded from the Block Ye to the more powerful Block I, allowing for the increased payload.

The spacecraft's testing was rigorous. A critical test of the parachute system on September 6, 1964, failed when the parachute hatch did not open. Engineers quickly identified the cause—an electrical circuit failure—and redesigned the system with redundancies. A subsequent test on October 5, 1964, was successful, as was an unmanned test flight on October 6, when the Voskhod capsule, renamed Kosmos-47, orbited the Earth for one day before safely returning under a parachute.

The development of the Voskhod program concluded with the successful launch of Voskhod 1 on October 12, 1964. Though the program was short-lived—Voskhod 2, flown in March 1965, was the final mission—it represented a significant achievement in the early years of human space exploration. The Voskhod program pushed the boundaries of what was possible, laying the groundwork for future advancements in multi-crewed space missions.

Voskhod 2

First extra-vehicular activity ("space walk")

Voskhod 2: The First Human Spacewalk and Its Aftermath

In March 1965, the Soviet Union launched Voskhod 2, a landmark mission that would etch its name in the annals of space exploration history. This mission was a follow-up to Voskhod 1, but it carried the ambitious goal of achieving the first extravehicular activity (EVA) or "spacewalk." Commanded by Pavel Belyayev and piloted by Alexei Leonov, Voskhod 2 was launched aboard a modified Vostok-based Voskhod 3KD spacecraft. This spacecraft, weighing 5,682 kg, featured significant adaptations, including an inflatable airlock—a crucial component that would enable Leonov to step outside the confines of the spacecraft. The mission would take humanity one step further into the unknown, demonstrating the feasibility of working outside a spacecraft in the vacuum of space.

Voskhod 2 launched on 18 March 1965 at 07:00 GMT. The spacecraft ascended into a highly elliptical orbit with an apogee of 475 km and a perigee of 167 km. The two-man crew prepared for the mission's main objective: Leonov's spacewalk. Approximately 90 minutes after entering orbit, Belyayev deployed and pressurized the Volga inflatable airlock, an essential spacecraft component that allowed Leonov to exit while maintaining cabin pressure for the internal systems safely. This innovation was required because Voskhod's avionics, which still used vacuum tubes, needed a stable atmosphere to function. Unlike the later American Gemini spacecraft, which used solid-state electronics and pure oxygen, Voskhod carried only limited reserves of nitrogen and oxygen, making an external airlock essential.

Once in orbit, Leonov donned the Berkut spacesuit, a modified version of the Vostok Sokol-1 suit, designed to support his activities outside the spacecraft. The suit was connected to an EVA backpack that provided 45 minutes of oxygen and regulated temperature by venting excess heat and moisture into space. At 08:34:51 GMT, Leonov began his spacewalk, stepping out into the cosmos as the first human to float freely outside a spacecraft. For the next 12 minutes, he orbited Earth tethered to the Voskhod by a 5.35-meter umbilical cord. Leonov's view from space was nothing short of breathtaking, allowing him to see from the Straits of Gibraltar to the Caspian Sea.

What began as a historic moment of triumph soon turned into a life-threatening situation. While Leonov performed his tasks—attaching a camera to the end of the airlock to record his EVA—his spacesuit began to balloon in the vacuum of space. The suit's pressurization caused it to stiffen, preventing him from bending his joints, and making it impossible for him to re-enter the airlock. To save himself, Leonov had to reduce the pressure in his suit below safety limits, risking decompression sickness, commonly known as the "bends." He did this without informing mission control to avoid alarming them. After struggling for several minutes, Leonov managed to force his body back into the airlock, but this maneuver was not without further complications. He entered the airlock head-first, violating procedure, and became stuck sideways. Once again, he was forced to reduce his suit pressure to free himself. Finally, at 08:47:00 GMT, Leonov re-entered the spacecraft.

The difficulties did not end with the EVA. The spacecraft's hatch suffered thermal distortion, complicating the process of sealing it properly. Worse still, a malfunction in the automatic landing system meant that the crew had to rely on manual re-entry procedures. Voskhod 2's cramped conditions—amplified by the astronauts wearing bulky spacesuits—made it difficult to return to their seats and stabilize the spacecraft's center of mass. This delay led to an inaccurate re-entry trajectory.

On 19 March 1965, Voskhod 2 re-entered Earth's atmosphere, but due to the 46-second delay in stabilizing the spacecraft, it landed 386 km off course, deep in the forests of the Ural Mountains in the Perm Krai region. Cut off from immediate rescue, Belyayev and Leonov found themselves stranded in an inhospitable landscape populated by bears and wolves. Despite the uncertainty, the cosmonauts remained calm. Both had experience with wilderness survival: Belyayev, a keen outdoorsman, and Leonov, who had spent time in the wild for artistic inspiration.

Soviet flight controllers were initially unaware of the crew's fate, but their families were informed that the cosmonauts had successfully landed. After four hours, a helicopter located the Voskhod capsule, but it was impossible to land in the densely forested area. Supplies, including warm clothing and provisions, were air-dropped to the crew, who spent a freezing night in the capsule, enduring temperatures as low as −5 °C. Although their survival kit included a TP-82 survival pistol, which became famous for being explicitly developed for future missions, it was the cosmonauts' endurance and training that saw them through the ordeal.

The next day, a rescue team arrived on skis, chopping wood and building a small log cabin to provide the cosmonauts with warmth and comfort. After a second night in the wilderness, Belyayev, Leonov, and the rescuers, skied several kilometers to the nearest suitable helicopter landing site. The crew was flown to Perm for initial medical checks and then to Baikonur for their mission debriefing.

Voskhod 2 was a monumental achievement, not just for the Soviet space program but for humanity's ongoing journey into space. Despite the numerous challenges faced during the mission—ranging from the first spacewalk to the harrowing survival story on Earth—it proved that humans could work in space, laying the groundwork for future extravehicular activities.

The Voskhod 2 capsule remains on display at the RKK Energiya Museum in Korolev, Russia.

Chapter 4 - The Gemini Missions and the Path to the Moon

Gemini 3

First piloted spacecraft orbit change

Gemini 3: Pioneering Maneuverability in Space

On March 23, 1965, NASA launched Gemini 3, the first crewed mission of Project Gemini, marking a significant step in America's journey to the Moon. Command Pilot Virgil I. "Gus" Grissom and Pilot John W. Young flew three low Earth orbits aboard a spacecraft they affectionately nicknamed Molly Brown. This mission was not only historic for being the first time two American astronauts flew together in space, but it also tested crucial spacecraft maneuverability, a vital component for future lunar missions.

Gemini 3's primary objective was to demonstrate the Gemini spacecraft's ability to adjust its orbit in space, a task that involved firing thrusters to change the size and shape of their trajectory. This was the first time an American spacecraft had executed such an in-space maneuver. The ability to adjust orbits would become critical for the success of rendezvous and docking procedures planned for the Apollo program's Moon landings. Grissom and Young's ability to shift their orbital plane and lower their altitude marked the mission as a significant milestone in NASA's spaceflight efforts.

The mission was also notable as the final crewed flight controlled from Cape Kennedy Air Force Station in Florida. Following this mission, all subsequent operations were handled from NASA's newly opened Manned Spacecraft Center in Houston, Texas, further symbolizing the growth and development of the U.S. space program.

Gus Grissom, a veteran of the Mercury program, made his second and final spaceflight as the Command Pilot of Gemini 3, while John Young made his spaceflight debut as Pilot. The mission's backup crew, Walter M. Schirra and Thomas P. Stafford would later serve as the prime crew for Gemini 6. The original crew selection for Gemini 3 included Alan B. Shepard as Command Pilot, but Shepard was grounded in late 1963 due to an inner ear disorder, which resulted in Grissom taking his place.

The support crew included Roger B. Chaffee and L. Gordon Cooper Jr., both serving as capsule communicators (CAPCOM) during the mission. Chaffee, stationed in Houston, would tragically later perish alongside Grissom in the Apollo 1 fire in 1967.

Gemini 3 launched atop a Titan II rocket from Cape Kennedy at 9:24 a.m. EST. Weighing 3,236.9 kg (7,136 lbs), the spacecraft orbited the Earth with a perigee of 161.2 kilometers (100.2 miles) and an apogee of 224.2 kilometers (139.3 miles). Its orbit inclined at 32.6 degrees, taking 88.3 minutes to complete each circuit.

One of the mission's most important objectives was to test the newly developed thruster system of the Gemini spacecraft, which allowed for adjustments in both altitude and orbital inclination. Shortly after the first orbit, over Corpus Christi, Texas, Grissom and Young fired the Orbit Attitude and Maneuvering System (OAMS) engines for 74 seconds, executing the first-ever crewed spacecraft orbital maneuver. The spacecraft's orbit changed from 161.2 x 224.2 kilometers to 158 x 169 kilometers, demonstrating Gemini's capacity for controlled maneuverability in space.

Additionally, the mission tested a communication system that had been designed for the cancelled Mercury-Atlas 10 mission. By injecting water into the plasma sheath that forms during re-entry, communications with ground control improved, marking a significant technical achievement.

Another first achieved during Gemini 3 was the controlled reentry, where the spacecraft produced aerodynamic lift, altering its reentry path. Though wind tunnel studies had predicted a certain degree of control during descent, the actual lift generated was less than anticipated, and Molly Brown missed its intended splashdown point by 84 kilometers (45 nautical miles). Despite this, the reentry successfully demonstrated the spacecraft's capability to adjust its descent profile.

Despite the mission's overall success, Gemini 3 was not without its challenges. Early in the flight, the astronauts noticed the spacecraft yawing slightly to the left, which they initially attributed to a stuck thruster. Upon further investigation, the issue was traced to a venting water boiler, but it posed no significant threat to the mission.

Two minor technical failures occurred during the mission. A lever essential to an experiment involving sea urchin eggs broke off, preventing the experiment from being completed. Additionally, a photographic objective was only partially successful due to an incorrect lens setting on the onboard camera.

One of the most memorable incidents of the flight involved a contraband corned beef sandwich that John Young had smuggled aboard. Though amusing to both astronauts, the sandwich posed a potential risk due to the loose crumbs that could interfere with the spacecraft's systems. The event prompted NASA to strictly enforce food safety protocols for all future missions.

After three orbits, the crew prepared for reentry. As Molly Brown descended toward Earth, Grissom and Young encountered an unexpected event: the spacecraft shifted from a vertical to a horizontal attitude, causing Grissom to hit his helmet's faceplate against the control panel, cracking the acrylic material. This prompted NASA to switch to stronger polycarbonate for future space helmets.

Upon splashdown in the Atlantic Ocean, Molly Brown landed 84 kilometers off target. Despite the miscalculation, the U.S. Coast Guard and the recovery ship, USS Intrepid, quickly retrieved the astronauts. Grissom and Young chose to remain inside the spacecraft until recovery, an understandable decision given Grissom's previous experience during his Mercury flight when Liberty Bell 7 sank shortly after splashdown.

Gemini 3 was a success in many respects, advancing NASA's understanding of space maneuverability and reentry dynamics. The mission was supported by over 10,000 personnel, 126 aircraft, and 27 ships, reflecting the vast logistical operation required to support America's space ambitions.

Mariner 4

First successful Mars flyby missiont

In the midst of the Space Race, as both the United States and the Soviet Union vied for supremacy in space exploration, NASA's Mariner program sought to push the boundaries of planetary science. Launched on November 28, 1964, Mariner 4 became a cornerstone of this effort, marking the first successful flyby of Mars and returning the first close-up images of the planet from deep space. This mission forever altered humanity's view of the Red Planet, shifting the scientific community's expectations about the possibility of life on Mars.

Mariner 4 was designed to conduct detailed scientific observations of Mars and to transmit these findings back to Earth. Its primary goal was to gather data on Mars' atmosphere, magnetic fields, and surface, while secondary objectives included measuring the characteristics of interplanetary space. At the time, hopes were high that Mars might harbor life or at least signs of past habitability, but the images returned by Mariner 4 showed a barren, cratered world more akin to the Moon than the Earth, fundamentally changing the perception of Mars as a "dead" planet.

Before this milestone could be achieved, however, the mission faced a significant setback. Mariner 4's twin spacecraft, Mariner 3, had failed just weeks earlier due to a malfunction with its payload shroud, which had not separated as planned after launch. Investigations at NASA's Jet Propulsion Laboratory (JPL) revealed that pressure differences during separation caused the failure, leading engineers to design a new, all-metal fairing. Despite the time pressure to meet the 1964 Mars launch window, the revised design was ready in time for Mariner 4's launch aboard an Atlas-Agena rocket from Cape Canaveral, Florida.

Once in space, Mariner 4 successfully deployed its solar panels and instruments, including a high-gain antenna and scientific equipment designed to study cosmic rays, solar wind, and charged particles in interplanetary space. The spacecraft's most significant feature was its television camera, which used a vidicon tube to capture images of the Martian surface. These images would be stored on a magnetic tape recorder and transmitted back to Earth.

Navigating in the vastness of space posed unique challenges for Mariner 4. The spacecraft relied on the star Canopus for orientation, as both Earth and Mars were too dim to serve as navigational reference points during much of the mission. Initially, the spacecraft struggled to acquire and maintain a lock on Canopus, often mistakenly locking onto other bright stars or stray light patterns. Eventually, engineers resolved the issue by adjusting the spacecraft's sensors, allowing the mission to proceed smoothly.

The spacecraft's journey to Mars took approximately seven and a half months, during which it performed a midcourse correction maneuver on December 5, 1964. This maneuver was critical to ensuring that Mariner 4 would pass close enough to Mars to achieve its scientific objectives. With precise adjustments, the spacecraft was on track for its historic encounter.

On July 14, 1965, Mariner 4 began its flyby of Mars. It passed within 9,846 kilometers (6,118 miles) of the Martian surface at a speed of 7 kilometers per second (4.3 miles per second). The onboard camera captured 21 images, covering about 1% of the planet's surface. These grainy, black-and-white images were transmitted to Earth over the following weeks, revealing a desolate landscape marked by craters and devoid of signs of life or liquid water. This discovery dashed many long-held hopes that Mars might be similar to Earth, or at least capable of supporting life.

Mariner 4's data transmission continued through August 1965, and the spacecraft remained operational until December 1967, when communication was finally lost. During its mission, Mariner 4 had provided invaluable insights into the Martian environment, measuring surface temperatures, atmospheric pressure, and the absence of a detectable magnetic field or radiation belts around Mars. These findings laid the groundwork for future missions and deepened scientific understanding of the Red Planet's history and geology.

Although the mission's data indicated that Mars was not the Earth-like world some had hoped for, Mariner 4 represented a major leap in space exploration. It was the first spacecraft to send back images from another planet, expanding humanity's reach beyond Earth's immediate neighborhood. In total, Mariner 4 returned 5.2 million bits of data, advancing the technology and methods that would be crucial for subsequent planetary exploration missions.

The mission cost $83.2 million—equivalent to approximately $804 million in 2023—and paved the way for the more advanced Mariner and Viking missions that would follow. In many ways, Mariner 4 marked the beginning of serious, systematic exploration of the planets, setting the stage for future missions that would continue to probe the mysteries of our solar system.

Gemini 6A & Gemini 7

First rendezvous of manned spacecraft

Gemini 6A: The First Crewed Rendezvous in Space

In December 1965, NASA's Gemini program made a historic leap with the Gemini 6A mission, which marked the first crewed rendezvous between two spacecraft. Commanded by astronaut Walter M. Schirra Jr., with Thomas P. Stafford as the pilot, Gemini 6A succeeded in a complex and delicate maneuver that brought it within mere feet of its sister spacecraft, Gemini 7. This mission showcased NASA's increasing capability to perform precise orbital operations, essential for future endeavors such as lunar landings.

Unlike earlier Soviet attempts with the Vostok program, which launched pairs of spacecraft that only established radio contact, Gemini 6A and Gemini 7 were able to adjust their orbits, coming as close as 30 centimeters—essentially within docking range, had their spacecraft been equipped for such a procedure. The rendezvous demonstrated the effectiveness of NASA's techniques for orbital navigation, a crucial milestone in the progression of human space exploration.

Gemini 6A was the fifth crewed mission in NASA's Gemini program and the 13th crewed American flight in space history. Launched on December 15, 1965, the spacecraft weighed 3,546 kilograms (7,818 pounds) and orbited the Earth with a perigee of 161 kilometers (100 miles) and an apogee of 259.4 kilometers (161.2 miles). The mission, which had a duration of five hours and 19 minutes, achieved its primary objective of rendezvousing with Gemini 7, flown by Frank Borman and James Lovell.

Although the Gemini 6A mission succeeded in rendezvousing with Gemini 7, it was originally conceived as Gemini 6, a mission that aimed to dock with an Agena Target Vehicle in October 1965. The original plan included multiple dockings with the Agena, allowing Schirra and Stafford to practice the maneuvers that would later become essential for the Apollo program's lunar docking missions.

The original Gemini 6 mission, scheduled for October 25, 1965, was designed to last 46 hours and 47 minutes, with the spacecraft completing 29 orbits of Earth. However, just 15 minutes before the launch, the Agena Target Vehicle, which was to serve as the docking target, was launched atop an Atlas booster. Although the Atlas booster successfully propelled the Agena into space, the Agena's engine malfunctioned shortly after ignition, causing a catastrophic failure. Telemetry was lost, and the vehicle was presumed to have exploded, scattering debris into the Atlantic Ocean. This led to the cancellation of the Gemini 6 launch.

With the Agena failure, NASA quickly adapted by revising the Gemini 6 mission into Gemini 6A. The new mission would no longer attempt a docking but instead focus on rendezvousing with Gemini 7, which was scheduled for a 14-day endurance mission. This revision called for Gemini 6A to launch on December 15, eight days after Gemini 7, and perform the world's first crewed orbital rendezvous. While there was discussion of performing an extravehicular activity (EVA) from Gemini 6A to Gemini 7, in which Stafford would swap places with Gemini 7's pilot Jim Lovell, this idea was abandoned due to concerns about Lovell needing to wear an EVA suit for an extended period.

Astronauts Thomas P. Stafford (left), pilot, and Walter
M. Schirra Jr., command pilot,

The Gemini 6A spacecraft launched without issue and quickly maneuvered to catch up to Gemini 7. After several orbital adjustments, Schirra expertly piloted Gemini 6A to within a foot of Gemini 7, allowing the two spacecraft to fly in tandem. The achievement was a testament to NASA's growing expertise in space operations, an essential capability for future Apollo missions that would require precise orbital rendezvous between the lunar module and command module.

The Gemini 6A mission was crewed by Schirra, a veteran of the Mercury program, and Stafford, who was making his first spaceflight. The backup crew consisted of Virgil "Gus" Grissom and John W. Young, both of whom had already flown in space. The support team included Charles Bassett, Alan Bean, Eugene Cernan, and Elliot See, all of whom served as Capsule Communicators (CAPCOMs), responsible for relaying mission control instructions to the astronauts.

On December 12, 1965, NASA made its first attempt to launch the Gemini 6A mission, an essential step in advancing space rendezvous techniques. Commanded by Wally Schirra, with Thomas P. Stafford as the pilot, Gemini 6A aimed to rendezvous with its sister spacecraft, Gemini 7. However, the launch attempt ended abruptly when the Titan II rocket engines ignited, only to shut down after 1.5 seconds. Despite mission rules dictating an immediate abort, Schirra's quick thinking prevented a dangerous ejection. Aware that the booster had not lifted off the pad, he refrained from pulling the D-ring that would have fired the ejection seats, a decision that likely saved both astronauts from injury or worse. The ejection seats, designed to thrust astronauts 800 feet into the air to avoid a potential explosion, were questionable in their reliability, especially given the high G-forces involved. Furthermore, the cabin had been soaking in pure oxygen, which posed a severe fire hazard had an ejection occurred.

Later, Stafford reflected on the situation, recalling that the cabin's oxygen saturation could have resulted in a fire similar to the tragic Apollo 1 incident that occurred years later. In a 1997 NASA oral history, he described the potential catastrophe, noting that "everything was soaked in oxygen." Fortunately, Schirra's decision to remain in the spacecraft averted this disaster.

Upon inspection, the failure was traced to a premature disconnection of an umbilical plug at the base of the rocket, which had signaled the spacecraft that liftoff had occurred. However, the Titan II booster had not yet achieved full thrust. Further investigation uncovered a secondary issue: a plastic dust cover had been left inside the gas generator, blocking the flow of oxidizer to one of the engines. With both problems identified and resolved, NASA cleared Gemini 6A for a second launch attempt.

On December 15, 1965, Gemini 6A successfully launched from Cape Kennedy's Launch Complex 19 at 8:37 a.m. EST. This marked the third attempt for Gemini spacecraft No. 6 and the second for the 6A mission. The launch proceeded smoothly, with the Titan II first stage cutting off at T+160 seconds and the second stage at T+341 seconds. Schirra and Stafford entered a 100-by-161-mile orbit, setting the stage for the planned rendezvous with Gemini 7.

The mission called for a rendezvous during the fourth orbit of Gemini 6A. Ninety-four minutes after launch, the crew performed their first burn, increasing their speed by 5 meters per second. This adjustment allowed them to gradually catch up to Gemini 7, which had launched eight days earlier on a 14-day endurance mission. At two hours and 18 minutes into the flight, Gemini 6A made a second burn to match Gemini 7's orbital inclination, closing the gap between the spacecraft to 300 miles.

Radar contact with Gemini 7 was established three hours and 15 minutes into the mission, when the two spacecraft were still 270 miles apart. A third burn placed Gemini 6A in a nearly identical orbit to Gemini 7's, allowing Schirra to begin the final approach. At five hours and four minutes, Schirra sighted what he first thought was the star Sirius, but it was actually Gemini 7, glowing brightly against the darkness of space.

After several precise burns, the two spacecraft came within 130 feet of each other. In the following hours, the two crews demonstrated the capabilities of NASA's rendezvous technology, flying as close as one foot apart. Schirra marveled at the ease with which he could maneuver in space, noting that the lack of turbulence allowed for smooth, confident movements. The two spacecraft remained in formation for several hours, with Schirra performing a series of fly-arounds and inspections of Gemini 7. He described the experience as "literally flying rings around it," maintaining perfect control as the spacecraft drew closer.

The mission's precision also proved efficient: despite the multiple burns, Gemini 6A had only used 112 pounds of fuel, leaving ample reserves for further maneuvers. Eventually, as the astronauts' sleep periods approached, Schirra and Stafford initiated a separation burn, drifting 30 kilometers away from Gemini 7 to prevent any accidental collisions during the night.

The next day, as the Gemini 6A crew prepared for reentry, they provided a lighthearted moment in what had been a serious and technically demanding mission. Schirra made a radio transmission to Mission Control, humorously claiming that they had spotted an unidentified flying object, resembling a satellite in a polar orbit. As the transmission continued, Schirra and Stafford produced small instruments: a harmonica and a set of bells, and proceeded to play "Jingle Bells." This was the first time musical instruments had been played in space, and the harmonica and bells were later displayed at the Smithsonian Institution.

On December 16, 1965, Gemini 6A fired its retrorockets, reentering the Earth's atmosphere and splashing down in the Atlantic Ocean, northeast of the Turks and Caicos Islands. The spacecraft landed within 18 kilometers of the planned recovery site and was picked up by the USS Wasp, an aircraft carrier equipped with a transportable satellite earth station. This marked the first time a spacecraft recovery had been televised live, thanks to an ITT relay via the Early Bird satellite.

The Gemini 6A mission, along with Gemini 7, was supported by over 10,000 Department of Defense personnel, 125 aircraft, and 16 ships. The successful rendezvous demonstrated NASA's growing mastery of space operations, essential for future missions to the Moon. It also solidified Schirra's reputation as one of NASA's most skilled pilots and showcased the flexibility and quick thinking of the astronauts involved in these groundbreaking missions.

Gemini 7: A Pioneering Mission in Long-Duration Spaceflight

In 1965, NASA's Gemini program embarked on a groundbreaking journey with the launch of Gemini 7. This mission marked the fourth crewed flight in the Gemini series and the twelfth crewed American spaceflight, solidifying the United States' role in the space race. Commanded by Frank Borman, with James A. Lovell serving as the pilot, the spacecraft carried its crew into orbit for nearly 14 days, setting new milestones in human space exploration.

Gemini 7's primary objective was to test the effects of long-duration spaceflight on the human body, as NASA prepared for future lunar missions. During their time in space, Borman and Lovell orbited Earth 206 times, spending a total of 330 hours in space—an endurance record that would stand until the Soviet Soyuz 9 mission in 1970. The flight also became the passive target for the first crewed space rendezvous, conducted by Gemini 6A, marking a significant moment in the evolution of space mission design.

Frank Borman and Jim Lovell, both on their first spaceflights, were backed by a capable support crew. Edward H. White II and Michael Collins were on standby as the backup crew, while Charles Bassett, Alan Bean, Eugene Cernan, and Elliot See served as capsule communicators (CAPCOM), ensuring seamless communication between the astronauts and mission control.

In the months leading up to launch, the Gemini 7 crew underwent rigorous training, including simulations of the challenging maneuvers required for long-duration space missions and rendezvous. The Gemini 7 mission was initially scheduled to follow Gemini 6, but a failed launch of the Agena Target Vehicle—intended for rendezvous with Gemini 6—altered the mission timeline. As a result, NASA opted to launch Gemini 7 as a rendezvous target for the hastily planned Gemini 6A mission, which would follow three days later.

Gemini 7 lifted off from Cape Kennedy on December 4, 1965, atop a Titan II rocket. The launch proceeded smoothly, with only minor pogo vibrations detected in the later stages of ascent. Once in orbit, the crew performed a series of initial maneuvers, including station-keeping with the spent upper stage of their rocket—a task that proved fuel-intensive and was eventually abandoned after 15 minutes.

From the outset, the mission was designed to address key challenges of long-term space habitation, such as waste management and crew comfort. NASA's ground control had scheduled the astronauts' work and sleep cycles to coincide with those of the prime shift ground crews. Unlike previous missions, both astronauts were allowed to sleep at the same time. They brought books along, as suggested by astronaut Pete Conrad, to combat the tedium of nearly two weeks in space.

As the days progressed, Borman and Lovell conducted a total of 20 experiments, more than on any prior Gemini mission. These studies included investigations into nutrition and physiological responses to the extended time in a weightless environment. They also tested a new lightweight spacesuit, the G5C, which proved uncomfortable in the spacecraft's confined quarters, leading the crew to remove their suits, a first in U.S. spaceflight eventually.

Throughout the mission, the crew grappled with the realities of extended space travel. Managing waste was one of the more challenging aspects, as the cramped spacecraft left little room for effective stowage. Despite the difficulties, the crew's meticulous attention to personal hygiene—having showered with anti-dandruff shampoo for weeks before the flight—prevented the buildup of skin flakes that had troubled previous missions.

On December 15, after overcoming a three-day delay due to a launch malfunction, Gemini 6A lifted off to begin its pursuit of Gemini 7. The mission's success hinged on a series of precise orbital maneuvers to align the two spacecraft. Walter Schirra, commanding Gemini 6A, executed the first burn 94 minutes after launch, followed by a phase adjustment that put his spacecraft on the same orbital inclination as Gemini 7.

As the two vehicles closed the gap, radar contact was established at 234 nautical miles. By the time the rendezvous was complete, the spacecraft were separated by a mere 130 feet, marking a historic moment in space navigation. For 270 minutes, the two crews communicated over radio, occasionally bringing the spacecraft within 1 foot of each other.

This successful rendezvous, achieved with minimal fuel consumption, demonstrated NASA's ability to conduct complex orbital mechanics, an essential skill for future lunar missions. As the astronauts approached their sleep period, Gemini 6A performed a separation burn to drift safely away from Gemini 7, avoiding any risk of collision during the night.

As Gemini 7 entered its final days in orbit, the excitement of the rendezvous gave way to the monotony of space travel. Both astronauts passed the time reading—Borman with Roughing It by Mark Twain, and Lovell with Drums Along the Mohawk by Walter D. Edmonds. However, as they entered their twelfth day in space, technical problems began to surface. Several thrusters malfunctioned due to outdated laminate in the thrust chambers, and the fuel cells provided only partial power. Nevertheless, the ship's batteries were sufficient to sustain them until reentry.

On December 18, after nearly 14 days in space, Borman and Lovell prepared for reentry. The retro-rockets fired flawlessly, guiding them back into Earth's atmosphere. They splashed down in the Atlantic Ocean, just 6.4 nautical miles from the planned recovery site. Despite the physical toll of prolonged weightlessness, both astronauts were in good health and spirits, joking with recovery crews aboard the USS Wasp about getting married after spending so much time together.

The Gemini 7 mission, supported by over 10,000 personnel, 125 aircraft, and 16 ships from the U.S. Department of Defense, was a resounding success. It not only tested the limits of human endurance in space but also demonstrated NASA's growing capability to perform complex orbital rendezvous, paving the way for future lunar missions in the Apollo program.

Luna 9

First soft landing on another celestial body (the Moon)
First photos from another celestial body

On February 3, 1966, the Soviet Union achieved a major milestone in space exploration with the success of Luna 9 (Луна-9), internally designated Ye-6 No.13, as part of its Luna program. Luna 9 became the first spacecraft to execute a controlled landing on a celestial body and transmit imagery from the surface. This accomplishment marked a significant victory in the ongoing space race between the Soviet Union and the United States, providing crucial insights into the lunar surface and paving the way for future lunar exploration.

The Luna 9 spacecraft consisted of a landing capsule and its carrier, together weighing 1,538 kilograms (3,391 lb) and standing 2.7 meters tall. Its design was crafted to endure the rigors of space travel and the high-speed descent onto the Moon. As the spacecraft neared the lunar surface, the lander capsule, weighing 99 kilograms (218 lb), was ejected to make a soft landing. This capsule, known as the Automatic Lunar Station (ALS), was a spheroid structure measuring 58 centimeters (23 inches) in diameter. It was equipped with a landing bag designed to absorb the impact of landing, which occurred at a speed exceeding 54 kilometers per hour (34 mph).

The capsule was hermetically sealed to protect its delicate equipment, including radio systems, heat control mechanisms, scientific instruments, power sources, and a television system intended to capture and relay images of the Moon's surface. Developed under the leadership of Sergei Korolev at the design bureau OKB-1, Luna 9 represented a leap forward in Soviet lunar efforts. Although Korolev had passed away shortly before the mission, his groundbreaking work continued under the Lavochkin design bureau, which took over the project as OKB-1 shifted focus toward human expeditions to the Moon.

After a series of unsuccessful Luna missions, Luna 9 became the first successful deep space probe built by Lavochkin, a design bureau that would later be responsible for nearly all Soviet (and later Russian) lunar and interplanetary spacecraft.

Luna 9 was launched atop a Molniya-M rocket, serial number 103-32, from Site 31/6 at the Baikonur Cosmodrome in the Kazakh Soviet Socialist Republic. Liftoff occurred at 11:41:37 GMT on January 31, 1966. The first three stages of the rocket successfully placed the spacecraft into low Earth orbit, with an altitude ranging from 168 to 219 kilometers (104 to 136 miles) and an inclination of 51.8 degrees. The fourth stage, a Blok-L engine, fired to boost the spacecraft into a highly elliptical geocentric orbit, with its apogee reaching approximately 500,000 kilometers (310,000 miles).

The spacecraft was spun at 0.67 rpm using nitrogen jets to maintain thermal balance. On February 1, a 48-second burn executed a mid-course correction, adjusting the spacecraft's trajectory by 71.2 meters per second (234 ft/s) and setting it on course for the Moon.

At an altitude of 8,300 kilometers (5,200 miles) from the Moon, Luna 9 prepared for its landing sequence. The spacecraft's spin was halted, and it oriented itself for the descent. Using an optomechanical system, it locked onto the Sun and Earth to ensure precise navigation. Upon reaching an altitude of 75 kilometers (47 miles) above the lunar surface, the spacecraft's radar altimeter activated, initiating the jettisoning of its side modules and inflating the airbags that would cushion the landing. The retro rockets fired, gradually reducing the spacecraft's speed.

At 250 meters (820 feet) above the lunar surface, the main retrorocket shut off, having reached the planned velocity for landing. Four smaller engines slowed the descent further. As the capsule neared the surface, a contact sensor triggered the shutdown of the engines, and the capsule was ejected from the main body. It landed safely at a speed of 22 kilometers per hour (14 mph) in the region known as Oceanus Procellarum, near the Reiner and Marius craters, at coordinates 7.08 N, 64.37 W. The capsule bounced several times before coming to a stop at 18:45:30 GMT on February 3, 1966.

Just over four minutes after landing, the four petals that enclosed the spacecraft's upper half unfolded, providing stability and exposing the scientific instruments and cameras. Seven hours later, once the Sun had risen to an elevation of 7 degrees, Luna 9 began transmitting the first images from the Moon's surface. These images included nine individual frames, five of which were panoramic shots showing the immediate surroundings, including rocks and the lunar horizon about 1.4 kilometers (0.87 miles) away.

Over the course of seven radio sessions, Luna 9 transmitted 8 hours and 5 minutes of data back to Earth, including three sets of television pictures. Soviet authorities did not immediately release the images, but scientists at Jodrell Bank Observatory in England, monitoring the spacecraft's signals, quickly noticed that the format used to transmit the pictures was identical to Radiofax, a system used by newspapers. The Daily Express rushed a receiver to the observatory, allowing the images to be decoded and published worldwide. This unexpected revelation provided the global public with the first-ever images from the surface of another celestial body.

In addition to capturing images, Luna 9 carried a radiation detector, which measured radiation levels on the Moon's surface. The instrument recorded a dosage of 30 millirads per day (0.3 milligrays), providing essential data about the lunar environment. One of the mission's key findings was that the lunar surface was solid enough to support the weight of a spacecraft, dispelling previous concerns that landers might sink into the lunar dust.

Luna 9 continued transmitting until February 6, 1966, when contact was lost at 22:55 GMT. By then, the mission had achieved its primary objectives, marking a turning point in the exploration of the Moon and contributing to the scientific foundation for future missions, both robotic and crewed.

Venera 3

First hard landing on another planet (Venus)

On November 16, 1965, the Soviet Union launched Venera 3 (Russian: Венера-3), a space probe from the ambitious Venera program, designed to explore the surface of Venus. Launched at 04:19 UTC from Baikonur, Kazakhstan, the spacecraft comprised two main components: an entry probe intended to descend through the Venusian atmosphere and land on the surface, and a carrier/flyby spacecraft that transported the entry probe toward Venus while collecting scientific data along the journey. The mission sought to achieve the first successful landing on another planet.

At the time, the Soviet Union's planetary exploration program had faced numerous challenges. Although Luna 2 and Luna 3 had succeeded in some of their objectives in the late 1950s, many Soviet missions had failed. Meanwhile, the United States had achieved notable success with its Mariner 2 Venus probe, which flew by the planet, and the Mariner 4 mission to Mars, which transmitted groundbreaking images of the Martian surface. In response to these challenges, the Soviet Union reassigned the planetary probe program in 1965 to the Lavochkin Bureau, replacing Sergei Korolev's OKB-1 design bureau.

The Lavochkin Bureau undertook a rigorous testing regime for both the Venera and Luna probes, something Korolev had traditionally resisted except for crewed spacecraft. This testing revealed significant design flaws, including the failure of the Venera landers to withstand the G-forces expected during their missions. This comprehensive overhaul was critical in preparing the Venera 3 probe for its journey to Venus.

The mission's entry probe was equipped with a range of scientific instruments, including a radio communication system, power sources, and medallions bearing the Coat of Arms of the Soviet Union. Its primary scientific objectives were to measure atmospheric conditions, study cosmic radiation, and analyze the surface of Venus. Notably, the probe's carrier spacecraft marked the first operational use of Gallium Arsenide (GaAs) solar cells in space, chosen for their superior performance in high-temperature environments. Two solar panels, each two square meters in size, powered the carrier spacecraft, while the entry probe used non-rechargeable batteries.

After launch, Venera 3 initially missed its intended trajectory, bypassing Venus by 60,550 kilometers. On December 26, 1965, a critical course correction brought the probe onto a collision path with the planet. However, contact with the probe was lost on February 15, 1966, likely due to overheating of the spacecraft's systems as it neared Venus. Despite this, Venera 3 became the first human-made object to impact another planet when it crash-landed on Venus on March 1, 1966.

The probe carried a suite of scientific instruments, including magnetometers for studying interplanetary magnetic fields, cosmic ray detectors, and piezoelectric sensors for micrometeoroid research. It also featured equipment to measure cosmic radio emissions and solar plasma flows. Although it did not carry a micrometeorite detector, it was equipped with a photometer, gas analyzer, and sensors to measure temperature, pressure, and density within Venus' atmosphere. These tools were intended to offer humanity its first direct insights into the environmental conditions on Venus.

Though Venera 3 did not transmit data back to Earth from the surface of Venus, its successful impact marked a significant milestone in planetary exploration, proving that reaching the surface of another planet was possible. The mission's legacy paved the way for future missions to Venus, contributing to our understanding of planetary environments and the challenges of interplanetary exploration.

Gemini 8 / ATV

First spacecraft docking

On March 16, 1966, NASA launched Gemini 8, the sixth crewed spaceflight of the Gemini program, marking a significant milestone in space exploration history. Commanded by Neil A. Armstrong and piloted by David R. Scott, this mission was designed to achieve the first docking of two spacecraft in orbit. Though the mission began with the promise of a groundbreaking achievement, it quickly turned perilous due to the first critical in-space system failure of a U.S. spacecraft, which threatened the lives of the astronauts and required an emergency mission abort. Despite the danger, Armstrong and Scott safely returned to Earth, demonstrating the crew's expertise in handling life-threatening situations in space.

The selection of Neil Armstrong as command pilot for Gemini 8 was a historic moment. Armstrong, having resigned his commission in the U.S. Naval Reserve in 1960, was selected as a crew member in September 1965. His flight on Gemini 8 marked only the second time a U.S. civilian flew into space, following Joe Walker's X-15 Flight 90, and the first time a U.S. civilian entered orbit. Armstrong, a highly skilled test pilot, was paired with David R. Scott, who was making his first spaceflight.

The backup crew consisted of Charles "Pete" Conrad Jr. as command pilot and Richard F. Gordon Jr. as pilot, both of whom would later fly the Gemini 11 mission. The support crew included notable astronauts Walter Cunningham as Cape CAPCOM and James A. Lovell as Houston CAPCOM, key figures in mission operations.

(Nov. 4, 1965) Astronauts David R. Scott (left), Pilot; and, Neil A. Armstrong (right), Command Pilot,

The primary objective of Gemini 8 was to rendezvous and dock with the Agena Target Vehicle (ATV), a feat that would represent the first docking of two spacecraft in orbit. Launched into a near-circular orbit of 161 nautical miles (298 kilometers), the Agena was prepped for this historic docking maneuver. The Gemini spacecraft, carrying Armstrong and Scott, was launched later the same day aboard a Titan II rocket, reaching orbit at approximately 10:41 a.m. EST.

Gemini 8 was planned as a three-day mission, during which Scott would perform a two-hour, 10-minute extravehicular activity (EVA), the first since Ed White's spacewalk on Gemini 4 in June 1965. Scott's ambitious EVA would involve retrieving scientific instruments and testing new tools and equipment, including a minimum-reaction power tool and an Extravehicular Support Pack (ESP) that would allow extended maneuverability in space. In addition to the EVA, the mission carried several scientific, technological, and medical experiments aimed at furthering knowledge of spaceflight dynamics and human adaptation to space.

The mission began with the launch of the Agena Target Vehicle from Cape Kennedy, Florida, aboard an Atlas rocket. The Agena successfully reached its designated orbit and aligned itself for docking. Five months earlier, an earlier launch of an Agena vehicle for Gemini 6 had failed when the vehicle's engine exploded during orbital insertion. This time, however, the launch proceeded flawlessly, putting the Agena in a stable orbit.

At 10:41 a.m. EST, Gemini 8 launched from Cape Kennedy aboard a Titan II rocket, coinciding with the 40th anniversary of Dr. Robert H. Goddard's historic launch of the world's first liquid-fueled rocket. The Gemini spacecraft was placed into an 86-by-147-nautical-mile (159 by 272 km) orbit, with all systems functioning nominally.

After a series of precise maneuvers, including adjustments to the spacecraft's apogee and perigee, Armstrong and Scott approached the Agena. They executed a series of burns to align the spacecraft with the target vehicle, with Armstrong carefully controlling their approach. At 3 hours, 48 minutes into the mission, they spotted the Agena and performed final adjustments, docking successfully at 23:14 UTC. Scott's excited report to Mission Control—"Flight, we are docked! Yes, it's really a smoothie"—marked a triumphant moment in space exploration.

Shortly after the docking, what should have been a historic moment turned into a life-threatening emergency. Soon after the Agena began executing a stored command program, Scott noticed that the docked spacecraft was rolling. Armstrong immediately used the Gemini's Orbit Attitude and Maneuvering System (OAMS) thrusters to stop the roll, but it quickly restarted. The astronauts realized they were dealing with a severe malfunction, likely within their own spacecraft.

With communication temporarily lost due to the spacecraft being out of range, Armstrong and Scott had to make critical decisions without guidance from Mission Control. As the rate of rotation increased, threatening both spacecraft, Armstrong undocked from the Agena. However, without the added mass of the Agena, the rotation of the Gemini spacecraft accelerated dramatically, reaching a dangerous 296 degrees per second. Armstrong took swift action, shutting down the OAMS and activating the Reentry Control System (RCS) thrusters. His calm and decisive response stabilized the spacecraft and prevented a catastrophic failure.

When communications resumed, the astronauts confirmed that a malfunctioning OAMS thruster had caused the issue. Nearly 75% of their reentry fuel had been expended in resolving the problem, leaving no choice but to abort the mission.

Gemini 8 reentered Earth's atmosphere over China, well ahead of its original schedule. The spacecraft splashed down safely in the Pacific Ocean, 500 miles east of Okinawa, Japan, where U.S. Air Force pilot Les Schneider quickly spotted it. A recovery team, led by the USS Leonard F. Mason, secured the capsule and brought Armstrong and Scott on board. Both astronauts were exhausted but in good condition, despite having suffered from seasickness after splashdown.

The recovery of Gemini 8 was swift, and within hours the spacecraft and its crew were on their way back to the United States. NASA officials, including fellow astronaut Walter Schirra, greeted the crew in Okinawa, before they were flown to Florida for medical tests and debriefing.

The subsequent investigation into the thruster malfunction revealed that the most likely cause was an electrical short, possibly due to static discharge. Engineers concluded that the malfunctioning thruster had continued firing even when switched off. Spacecraft designs were modified to include isolated circuits for each thruster to prevent future occurrences.

NASA also formalized its investigation procedures following the incident. Dr. Robert Seamans, NASA's Deputy Administrator, established new guidelines for handling mission failures, later invoked during critical events like the Apollo 1 fire and the Apollo 13 mission.

Luna 10

First artificial satellite to orbit another celestial body (the Moon)

On March 31, 1966, the Soviet Union launched Luna 10 (also known as Lunik 10), a robotic spacecraft as part of the ambitious Luna program. It became the first artificial satellite of the Moon, marking a significant milestone in the history of space exploration. This groundbreaking mission conducted extensive research in lunar orbit, contributing crucial data about the Moon and its environment.

Luna 10 carried a range of scientific instruments designed to study the Moon's magnetic field, radiation belts, and the nature of its surface. One of the mission's most important findings was the detection of mass concentrations, or "mascons," which are areas of high density beneath the lunar mare basins that cause distortions in the orbits of spacecraft. Although the discovery of mascons is often credited to the American Lunar Orbiter missions, Luna 10 provided the first evidence of their existence.

Part of the E-6S series, Luna 10 had an on-orbit dry mass of 540 kilograms and was powered by batteries. Its suite of scientific instruments included a gamma-ray spectrometer, which measured energies between 0.3 and 3 MeV, a triaxial magnetometer to study the Moon's magnetic field, a meteorite detector, instruments for studying solar plasma, and devices to measure infrared emissions from the Moon. In addition, it conducted gravitational studies and cosmic radiation measurements, providing valuable data about the Moon's environment.

The mission's spacecraft launched on March 31, 1966, at 10:48 GMT. After a successful midcourse correction on April 1, Luna 10 entered lunar orbit on April 3 and completed its first orbit three hours later. The spacecraft's main bus carried a 245-kilogram instrument compartment, which separated after achieving an elliptical orbit of 350 x 1,000 kilometers, inclined at 71.9° to the lunar equator. Throughout the mission, Luna 10 completed 460 orbits around the Moon and transmitted 219 sets of data to Earth before radio signals were discontinued on May 30, 1966.

In addition to its scientific achievements, Luna 10 had a symbolic mission. The spacecraft carried solid-state oscillators programmed to play the notes of The Internationale, a revolutionary anthem associated with socialist movements. The intent was to broadcast this music to the 23rd Congress of the Communist Party of the Soviet Union. During a rehearsal on the night of April 3, the playback of The Internationale went as planned. However, a missing note in the transmission was discovered the following morning. To avoid embarrassment, controllers played the previous night's rehearsal tape to the assembled party officials, claiming it was a live broadcast from the Moon.

Despite the minor technical glitch, Luna 10 made history as the first artificial satellite of the Moon. The mission's scientific findings, particularly its contributions to the understanding of lunar mascons and the Moon's magnetic environment, were significant steps forward in lunar science and exploration. Luna 10 remains a testament to the early achievements of the Soviet space program during the height of the space race.

Gemini 11 / ATV

First direct-ascent (first orbit) rendezvous

Gemini 11, the ninth crewed mission of NASA's Project Gemini, marked a series of significant accomplishments during its flight from September 12 to 15, 1966. Commanded by Charles "Pete" Conrad Jr., with Richard F. Gordon Jr. as pilot, the mission was a crucial step in advancing NASA's capabilities in space rendezvous, docking, and extra-vehicular activities (EVAs), all vital for the upcoming Apollo lunar missions. Gemini 11 set new records and conducted scientific experiments that pushed the boundaries of human space exploration.

Gemini 11's primary objective was to perform a direct-ascent rendezvous with an Agena Target Vehicle, simulating a Lunar Module rendezvous with the Command/Service Module after a lunar landing. Achieving this, Conrad and Gordon docked with the Agena vehicle just 1 hour and 34 minutes after launch, an unprecedented feat that relied heavily on the spacecraft's onboard computer and radar systems, with minimal assistance from ground control. This success demonstrated the precision necessary for future lunar missions.

Astronauts Charles Conrad Jr. (right), command pilot,
and Richard F. Gordon JR., pilot

Once docked, Gemini 11 used the Agena's rocket engine to raise its apogee to a record-breaking 853 miles (1,373 kilometers), the highest Earth orbit ever reached by a crewed spacecraft until SpaceX's Polaris Dawn mission in 2024. This high-apogee elliptical orbit allowed the astronauts to conduct a variety of scientific experiments and demonstrate the potential for reaching greater distances, crucial for lunar and planetary exploration. Although later Apollo missions traveled farther to the Moon, the altitude achieved by Gemini 11 remained a benchmark for Earth orbit until the 21st century.

Another key objective was to test passive attitude stabilization by tethering the Gemini spacecraft to the Agena and creating artificial gravity. Conrad and Gordon extended a 100-foot (30-meter) tether between the two spacecraft and initiated a slow rotation by firing thrusters, generating a small but measurable artificial gravity of about 0.00015 g. Although the tether experiment had some challenges—such as difficulty keeping the tether taut—it provided valuable data for future missions considering similar techniques.

Gemini 11 included two EVAs, both performed by Richard Gordon. The first, on September 13, 1966, was scheduled to last two hours but was cut short due to the extreme physical demands of working in the vacuum of space. Gordon spent 33 minutes outside the spacecraft, attaching the tether to the Agena for the stabilization experiment. As with previous Gemini missions, performing extended work during an EVA proved more fatiguing than anticipated, reflecting the need to improve EVA suit design and astronaut training.

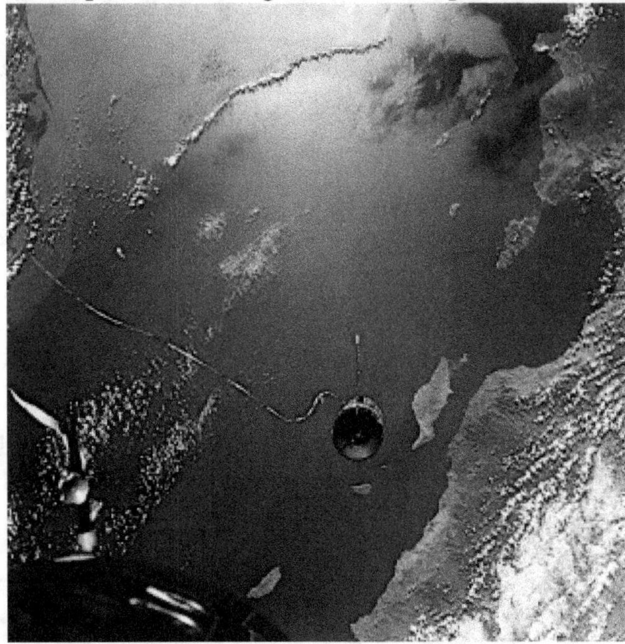

The Agena Target Docking Vehicle is tethered to the Gemini 11 spacecraft during its 31st revolution of the earth. Area below is the Gulf of California and Baja California at La Paz

Gordon's second EVA, on September 14, was a "stand-up" EVA, where he remained partially inside the spacecraft with his head and shoulders outside the hatch. This EVA lasted for over two hours and involved photographing the Earth, clouds, and stars. Unlike the first EVA, this was not physically taxing, and Gordon successfully completed his tasks, contributing to valuable scientific observations.

During their time in space, Conrad and Gordon conducted 12 scientific experiments spanning multiple disciplines. These experiments included:

Mass Determination: A technique to measure the mass of an orbiting object, in this case, the Agena, which had direct applications for future spacecraft design.

Synoptic Terrain and Weather Photography: High-quality photographs of Earth's surface and weather patterns were taken to support research in geology, oceanography, and meteorology.

Radiation and Zero-G Effects on Blood and Neurospora: This experiment studied the effects of space radiation on human blood cells and fungi, providing data on the biological impacts of space travel.

Power Tool Evaluation: An assessment of astronauts' ability to perform tasks in space using a power tool, which was critical for preparing future missions requiring complex in-space construction.

Other experiments, such as ultraviolet astronomical photography and studies of cosmic radiation, contributed to the growing body of knowledge about space physics and the Earth's atmospheric phenomena.

Gemini 11 concluded with NASA's first fully automated, computer-controlled reentry. This technological milestone brought the spacecraft down just 2.8 miles (4.5 kilometers) from the recovery ship, USS Guam, a mere 1.5 miles (2.4 kilometers) from the planned splashdown point. The precision of this reentry underscored the advances in NASA's navigation and computer systems.

Over 9,000 U.S. Department of Defense personnel, including 73 aircraft and 13 ships, supported the mission, demonstrating the extensive logistical coordination required for human spaceflight.

Gemini 11 was a resounding success, accomplishing all of its objectives and setting new records in space exploration. The mission provided critical insights into spacecraft rendezvous, docking, and EVAs, all of which would be essential for the success of the Apollo program. Conrad and Gordon's achievements helped solidify NASA's technical capabilities, proving that astronauts could perform complex tasks in space and return safely to Earth. This mission stands as a testament to the ingenuity and determination that propelled humanity closer to landing on the Moon.

Venera 4

First in situ analysis of the atmosphere of another planet (Venus)

Venera 4, designated 4V-1 No.310, was a significant milestone in the Soviet Union's Venera program, aimed at exploring Venus. Launched in June 1967, this probe was the first successful spacecraft to conduct in-situ analysis of another planet's atmosphere, providing groundbreaking data on Venus. Venera 4 consisted of two main components: a lander designed to enter Venus' atmosphere and parachute to the surface, and a carrier/flyby spacecraft that transported the lander and acted as a communication relay.

Venera 4 marked a major achievement in space exploration, offering the first chemical analysis of Venus' atmosphere, which revealed that it was composed primarily of carbon dioxide, with smaller amounts of nitrogen, oxygen, and water vapor. As it descended through the planet's thick atmosphere, Venera 4 became the first spacecraft to survive atmospheric entry on another planet. The probe detected a weak magnetic field and no radiation field, while discovering that the outer layers of Venus' atmosphere contained very little hydrogen and no atomic oxygen. The probe's direct measurements showed Venus to be far hotter and denser than expected, confirming that most of the planet's water had vanished long ago.

The Venera 4 mission required an innovative spacecraft to achieve these scientific objectives. The main carrier spacecraft stood 3.5 meters high, with solar panels spanning 4 meters and an area of 2.5 square meters. Equipped with instruments such as a magnetometer, ion detector, cosmic ray detector, and ultraviolet spectrometer, the carrier spacecraft was designed to operate until it released the lander. A liquid-fuel thruster allowed flight course adjustments, and the spacecraft could receive up to 127 different commands from Earth for course corrections.

The lander itself was a nearly spherical capsule, 1 meter in diameter and weighing 383 kilograms, designed to withstand the extreme conditions of Venus' atmosphere. The capsule featured an improved heat shield, capable of enduring temperatures up to 11,000°C, replacing earlier liquid-based cooling systems with a simpler gas cooling system. To ensure the lander's durability, it was subjected to rigorous testing, including exposure to high temperatures, pressure simulations up to 25 atmospheres, and accelerations of up to 450 g in a centrifuge. These tests revealed some issues, such as cracked electronic components, but repairs were made in time for the mission's tight launch window.

The capsule was designed to float in case of an unexpected water landing. Its parachute, able to withstand temperatures of 450°C, deployed during descent, releasing two transmitters that relayed atmospheric data back to Earth. The onboard instruments included a thermometer, barometer, hydrometer, and gas analysis equipment, sending data every 48 seconds.

To avoid biological contamination of Venus, the Venera 4 spacecraft was sterilized before launch. Additionally, the lander's battery, which powered the instruments and transmitters for up to 100 minutes, was recharged during the journey using the carrier spacecraft's solar panels.

Two probes were prepared for this mission: Venera 4, which successfully launched on June 12, 1967, and Kosmos 167, launched on June 17 but which failed to depart low Earth orbit. Venera 4 successfully reached Venus after a course correction on July 29, 1967, when it was 12 million kilometers from Earth. The probe entered Venus' atmosphere on October 18, 1967, after its 129-day journey, surviving the extreme heat and pressure during descent.

As the capsule descended, it deployed its parachute at an altitude of approximately 52 kilometers, transmitting data about the atmosphere's temperature, pressure, and composition. At 52 kilometers, the temperature was recorded at 33°C, with a pressure of less than 1 atmosphere. By the end of its descent, the temperature had reached 262°C, and the pressure had risen to 22 atmospheres, at which point signal transmission ceased. Venera 4 measured the atmospheric composition as 90-93% carbon dioxide, 0.4-0.8% oxygen, 7% nitrogen, and 0.1-1.6% water vapor.

The data transmitted by Venera 4 provided the first direct insight into Venus' atmosphere, correcting earlier assumptions about the planet's environment. The probe's findings revealed that Venus had an unexpectedly high concentration of carbon dioxide, a much denser atmosphere than anticipated, and no detectable radiation belts. The magnetic field measured by Venera 4 was 3,000 times weaker than Earth's, and the hydrogen corona around Venus was 1,000 times less dense than expected. These discoveries led to the conclusion that Venus had likely lost any water it once had long ago, a realization that challenged earlier beliefs due to thick clouds in the planet's atmosphere.

The mission was hailed as a success, especially given the previous failures of the Venera program. While Venera 4 was not designed to withstand the full pressure at Venus' surface, it paved the way for future missions. The first successful soft landing on Venus was achieved by Venera 7 in 1970, building on the technological advances and lessons learned from Venera 4's historic mission.

Venera 4 was a turning point in planetary exploration, demonstrating the feasibility of atmospheric entry and data collection from another planet. It provided vital information that helped refine models of Venus' atmosphere and contributed to a deeper understanding of planetary environments. The collaborative analysis of Venera 4 data and the American Mariner 5 mission showcased early international cooperation in space science, setting the stage for future missions and joint exploration efforts.

Cosmos 186 / Cosmos 188

First docking of two remote-controlled spacecraft

On October 27, 1967, the Soviet Union achieved a milestone in space exploration with the launch of Kosmos 186, a spacecraft designed as part of the Soyuz program. Kosmos 186, which carried a descent module for landing scientific instruments and test objects, was launched aboard a Soyuz 11A511 booster from Baikonur Cosmodrome's Site 31/6. Placed in a low Earth orbit with a perigee of 172 kilometers (107 miles) and an apogee of 212 kilometers (132 miles), the spacecraft's mission aimed to test a fully automated docking procedure—an ambitious objective given the limitations of the time.

The mission followed the tragedies of Soyuz 1 and NASA's Apollo 1 earlier in 1967, which led both space programs to adopt a cautious approach. For the Soviets, this meant launching uncrewed spacecraft to validate docking technologies critical for future crewed missions. However, because the Soviet Union lacked ground stations outside its territory, the docking had to be fully automated—a significant challenge.

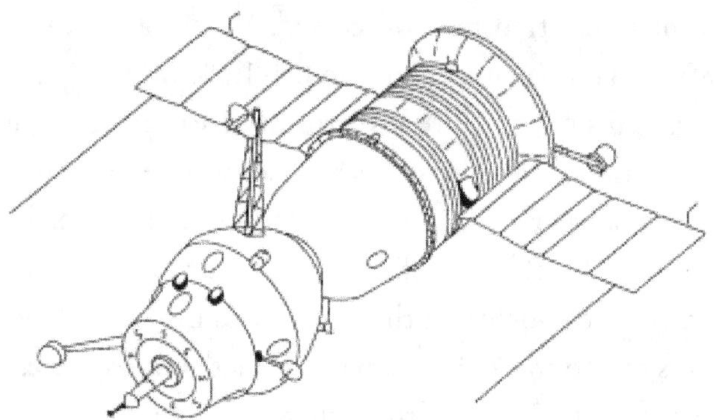

Three days later, on October 30, 1967, Kosmos 188 was launched from Baikonur Cosmodrome's Site 1/5 aboard another Soyuz 11A511 booster. This spacecraft, with an orbital perigee of 180 kilometers (110 miles) and an apogee of 247 kilometers (153 miles), served as the passive target for the docking attempt. The docking maneuver took place just 62 minutes after the launch of Kosmos 188, making history as the first fully automated docking between two spacecraft.

The docking was conducted using the IGLA system onboard Kosmos 186, which autonomously managed the mutual search, approach, mooring, and docking of the two spacecraft over the South Atlantic. Although the spacecraft successfully docked mechanically, the mission was not flawless. An electrical connection could not be established between the two vehicles, and the maneuver consumed more fuel than anticipated. Despite these challenges, the docking itself was a significant achievement.

After three and a half hours of joint flight, the spacecraft separated and continued to orbit independently. Kosmos 186 made a soft landing in a predetermined region of the Soviet Union on October 31, 1967, while Kosmos 188's descent module was destroyed when its emergency self-destruct mechanism accidentally triggered during reentry on November 2, 1967. Soviet engineer Boris Chertok, a key figure in the Soviet space program, later confirmed the destruction of Kosmos 188.

The mission, despite its anomalies, represented a critical step forward for Soviet space technology. It demonstrated that spacecraft could dock in orbit autonomously—a capability essential for future human spaceflight programs. While the Soviet lunar landing program was eventually canceled, the technologies developed for automatic docking were later employed in the successful Salyut and Mir space station programs. Today, these automated docking systems remain a core component of operations on the Russian segment of the International Space Station (ISS).

This mission also underscored the feasibility of assembling large space structures by docking smaller components in orbit, eliminating the need for larger rockets. This principle became a cornerstone of space station construction, a legacy that continues to influence modern space exploration efforts.

Zond 5

First return to Earth after circling the Moon
First life forms to circle the Moon (returned safely)

Zond 5, part of the Soviet Union's ambitious Zond program, achieved a historic milestone in September 1968. It became the first spacecraft to travel to and circle the Moon, completing a circumlunar trajectory and safely returning to Earth. This mission marked several key firsts: it was the first Moon mission to carry living organisms, including two Russian tortoises, and the first to return terrestrial life safely from the vicinity of the Moon. Zond 5's mission was a major step forward in the space race, demonstrating the Soviet Union's capacity to send spacecraft around the Moon and recover them, while also carrying out biological experiments on space travel's effects.

The Zond spacecraft was a modified version of the Soyuz 7K-L1, designed for crewed lunar flybys, but the mission was uncrewed due to concerns over previous mission failures. Earlier in 1968, both Zond 1968A and Zond 1968B missions ended in failure, with the latter resulting in a tragic explosion on the launchpad, killing three workers. These setbacks pushed the Soviet Union to take a more cautious approach, deciding against sending cosmonauts on Zond 5. Instead, animals and biological specimens would serve as test subjects, with two tortoises being the most prominent members of the payload.

Zond 5's biological payload included a diverse array of organisms. The tortoises, weighing between 0.34 and 0.4 kilograms, were placed in the spacecraft 12 days before launch and deprived of food and water, as part of an experiment to study the biological impacts of space travel. Alongside the tortoises, the spacecraft carried insects, fruit fly eggs, microorganisms, and various plant cells, including wheat, barley, pea, and pine. The purpose of this variety was to examine the effects of cosmic radiation on different life forms, although many of the chosen species were extremophiles, which could withstand higher radiation levels than humans.

The mission launched on September 14, 1968, from the Baikonur Cosmodrome, using a Proton-K carrier rocket with a Block D upper-stage. Zond 5 entered a parking orbit at an altitude of 191 by 219 kilometers before the upper-stage ignited to send the spacecraft on a trans-lunar injection. The mission faced early challenges when a contaminated star tracker caused issues with the spacecraft's attitude, delaying an essential course correction. However, by using the Sun and Earth as reference points, Zond 5 was able to correct its trajectory en route to the Moon.

On September 18, Zond 5 flew past the Moon at a closest distance of 1,950 kilometers. Although it did not enter lunar orbit, it provided a close approach, capturing valuable data. As the spacecraft returned to Earth, additional technical difficulties arose, including the failure of another star tracker and the erroneous shutdown of the guided reentry system. Despite these complications, Zond 5 reentered Earth's atmosphere on September 21, 1968, splashing down in the Indian Ocean, 105 kilometers from the nearest Soviet recovery ship. The recovery took place at night, delaying efforts, but the spacecraft and its biological specimens were ultimately retrieved.

The mission's biological results were carefully analyzed. The tortoises, called No. 22 and No. 37, had lost 10% of their body weight during the flight but showed no significant health issues. In fact, the post-flight analysis revealed that the biological changes observed in the tortoises, such as elevated iron and glycogen levels, were likely due to starvation rather than the effects of space travel. Comparisons with control tortoises, which were similarly deprived of food but remained on Earth, showed minimal differences, leading Soviet scientists to conclude that space travel had little effect on the animals' health.

Zond 5 also contributed to space exploration in other ways. The spacecraft took high-quality photographs of Earth from 90,000 kilometers, the first images of their kind, providing a new perspective of our planet. These photos, along with the mission's success in returning to Earth, bolstered Soviet claims that they were leading the space race. British astronomer Bernard Lovell remarked that the mission demonstrated Soviet technical superiority, suggesting that a crewed mission to the Moon could be imminent. However, American sources pointed out that Zond 5's mission had not been without flaws. The spacecraft had flown farther from the Moon than planned, limiting the usefulness of its lunar photography, and its steep reentry angle would have been dangerous for a cosmonaut.

Despite these critiques, the Soviets touted the mission as a precursor to a crewed lunar flight, with official announcements in November 1968 emphasizing the spacecraft's success in carrying living organisms around the Moon. This announcement came just a month before NASA's Apollo 8 mission, as the two superpowers raced to achieve the first crewed lunar mission.

One particularly dramatic episode during the mission involved a hoax staged by Soviet cosmonauts, which briefly led the world to believe that humans might be onboard Zond 5. On September 19, 1968, voices of cosmonauts Valery Bykovsky, Vitaly Sevastyanov, and Pavel Popovich were broadcast from the spacecraft and intercepted by the CIA and the Jodrell Bank Observatory. The transmissions, which included telemetry data and discussions about a potential lunar landing, caused a stir in the United States. It was later revealed that the transmissions were a prank orchestrated by the cosmonauts, who had no intention of landing on the Moon but aimed to rattle their American counterparts. Popovich would later recall the episode with amusement, noting how it briefly unsettled the U.S. space program.

Though a crewed lunar mission never materialized for the Soviet Union, the achievements of Zond 5 marked an important chapter in the space race, providing valuable insights into the biological effects of space travel and the challenges of deep-space exploration.

Chapter 5 - The Race to the Moon

Apollo 8

First return to Earth after orbiting the Moon First human spaceflight mission to enter the gravitational influence of another celestial body

During the 1960s, Project Apollo emerged as the centerpiece of NASA's efforts to land humans on the Moon, spurred by President John F. Kennedy's 1961 challenge to achieve this goal by the end of the decade. Central to the Apollo program was the decision to adopt the "lunar orbit rendezvous" strategy. This approach involved a specialized lunar module descending to the Moon's surface while the command module remained in lunar orbit. By focusing on a more efficient, lighter spacecraft, NASA could use the powerful Saturn V rocket to launch the mission. The command module, where the astronauts would live and work, was the only portion designed to return to Earth, while the service module provided vital life support and propulsion during the journey.

One of the most significant milestones in the Apollo program was Apollo 8, the first crewed mission to utilize the Saturn V rocket. The crew, comprising Commander Frank Borman, Command Module Pilot James Lovell, and Lunar Module Pilot William Anders, was officially announced on November 20, 1967. Originally, Michael Collins had been assigned as Command Module Pilot, but a cervical disc herniation led to Lovell's reassignment in July 1968. This mission was notable not only for being the first crewed flight of the Saturn V but also for its unique crew composition. Frank Borman, despite being the commander, was not the most experienced astronaut on the mission. Lovell, with two previous spaceflights under his belt (Gemini VII and Gemini XII), became the first astronaut to fly in a non-commander role after having previously commanded a mission.

The Apollo 8 mission was further distinguished by its backup crew, which included future Apollo 11 astronauts Neil Armstrong and Buzz Aldrin, along with Fred Haise, who would later play a pivotal role in the Apollo 13 mission. The assignment of backup crews ensured that trained personnel were ready to step in if needed, while also providing critical mission experience for future astronauts. Apollo missions operated with a three-tiered astronaut crew system: the prime crew, backup crew, and support crew. For Apollo 8, the support crew—Ken Mattingly, Vance Brand, and Gerald Carr—played a vital role in preparing mission-critical documents, such as flight plans and procedures, which were rehearsed by the prime and backup crews during rigorous simulator training.

Communication with the crew during the mission was handled by the Capsule Communicator (CAPCOM), the only person authorized to speak directly with the astronauts. Apollo 8's CAPCOM team consisted of experienced astronauts like Michael Collins, Neil Armstrong, and Buzz Aldrin, ensuring that the crew received instructions from individuals familiar with spaceflight challenges. On the ground, mission control teams worked in shifts to monitor every aspect of the flight. The mission's success relied on the leadership of Flight Directors Clifford E. Charlesworth, Glynn Lunney, and Milton Windler, who each commanded their teams during critical phases of the mission.

NASA's seven-step plan for the Apollo program, outlined on September 20, 1967, was designed to culminate in the historic Moon landing. Early missions tested various components essential to this goal. Apollo 4 and Apollo 6 were uncrewed tests of the Saturn V rocket and the command and service module (CSM), designated as "A" missions.

Apollo 5 tested the lunar module (LM) in Earth orbit as a "B" mission, and Apollo 7, launched in October 1968, was a crewed Earth-orbit flight of the CSM, classified as a "C" mission. As the lunar module faced production delays, the "D" mission, planned for December 1968, would have tested the LM in low Earth orbit with astronauts James McDivitt, David Scott, and Russell Schweickart aboard. Apollo 8, originally intended to be the "E" mission, was set to conduct a medium Earth orbit test of both the LM and CSM in early 1969, followed by the "F" mission, which would test the spacecraft in lunar orbit, and the "G" mission, the eventual Moon landing.

However, delays in the lunar module's development threatened this timeline. When LM-3 arrived at Kennedy Space Center in June 1968, engineers identified over 100 critical defects. The module's readiness was postponed to early 1969, jeopardizing NASA's ability to meet Kennedy's deadline. In response, George Low, Manager of the Apollo Spacecraft Program Office, proposed an innovative solution: to fly a command and service module-only mission to the Moon in December 1968. This mission, known as the "C-Prime" mission, would allow NASA to test crucial systems in lunar orbit without the lunar module, thus keeping the overall timeline intact.

On August 9, 1968, Low presented his proposal to key NASA figures, including Bob Gilruth, Director of the Manned Spacecraft Center; Flight Director Chris Kraft; and Director of Flight Crew Operations Donald Slayton. Afterward, they traveled to the Marshall Space Flight Center in Huntsville, Alabama, to discuss the idea with Apollo Program Director Samuel Phillips, KSC Director Kurt Debus, and Wernher von Braun, the head of the Saturn V rocket program. Following a thorough review of the technical and operational risks, NASA's senior leadership unanimously supported the C-Prime mission. Confidence in the spacecraft and personnel was high, and the Saturn V rocket, designated AS-503, was deemed ready by December 1, having resolved the pogo oscillation issues that affected Apollo 6.

With final approval from NASA Administrator James Webb, Apollo 8 was officially greenlit as a lunar orbit mission. This ambitious decision allowed NASA to maintain its momentum toward achieving a lunar landing by the end of the decade, providing a vital stepping stone for the eventual success of Apollo 11.

The decision to fly Apollo 8 as a lunar orbit mission marked a pivotal shift in NASA's strategy, one that dramatically altered crew assignments. Initially, James McDivitt's crew had been preparing for a critical lunar module test, but when it became clear that the lunar module would not be ready in time for the mission, McDivitt's team chose to remain focused on their task. Consequently, Frank Borman's crew, which had originally been slated for the "E" mission, was reassigned to fly Apollo 8. This adjustment required Borman's team to use CSM-103, the command module next in line for flight, as the lunar module (LM-3) would not be ready. David Scott, who had supervised the development of CSM-103, initially had reservations about this change, but the crew ultimately accepted the mission, understanding the urgency of NASA's lunar goals.

Left to right: Lovell, Anders, Borman

NASA's decision to proceed with Apollo 8 as a lunar orbit mission was driven not only by internal factors but also by the intensifying space race with the Soviet Union. In September 1968, the Soviets successfully launched Zond 5, an uncrewed spacecraft carrying living creatures, including Russian tortoises, on a circumlunar trajectory. This mission, which returned safely to Earth, raised alarm at NASA that the Soviets might soon send cosmonauts on a similar mission, potentially overshadowing Apollo 8. Adding to NASA's concerns, American reconnaissance satellites detected a mock-up of the N1 rocket—the Soviet counterpart to the Saturn V—on a launch pad at Baikonur in late 1967. These developments heightened the pressure on NASA to maintain its lead in the race to the Moon.

Despite these geopolitical pressures, NASA remained focused on ensuring the technical readiness of Apollo 8's launch vehicle, the Saturn V rocket designated AS-503. Originally assembled for an uncrewed Earth-orbit test, AS-503 had undergone significant modifications following the problematic Apollo 6 flight in April 1968. Severe pogo oscillations, engine failures, and a malfunctioning third stage had plagued that mission. Engineers at the Marshall Space Flight Center (MSFC) worked intensively to resolve these issues in preparation for Apollo 8. The pogo oscillation, which caused dangerous vibrations throughout the rocket, was addressed by installing helium gas dampeners designed to absorb the resonance. Engine failures, traced to ruptured fuel lines and miswiring, were corrected in subsequent redesigns, making the rocket more reliable for crewed missions.

By August 1968, NASA had rectified these problems and decided to use AS-503 for the crewed lunar mission. The rocket was fully assembled in the Vehicle Assembly Building on September 21, 1968, and made its slow, 3-mile journey to Launch Pad 39A atop a massive crawler-transporter on October 9. As Apollo 8's December launch date approached, testing continued at a rapid pace. The final readiness checks were conducted just three days before the scheduled launch, ensuring that all systems were prepared for the historic mission.

Apollo 8's success would hinge on the modifications made to the Saturn V and the lessons learned from Apollo 6. By the time the spacecraft was mounted atop the rocket, NASA had conducted extensive tests to verify the rocket's reliability. Mid-December saw the final round of system checks, including critical tests of the mechanisms designed to prevent pogo oscillation and engine malfunctions. Confident in the modifications, NASA was ready to embark on the boldest human spaceflight mission to date—a journey around the Moon and back. Apollo 8's success not only paved the way for the first lunar landing in 1969 but also reaffirmed NASA's technological leadership during a pivotal moment in the space race.

In the days leading up to the launch, Apollo 8's crew had a special visit from aviation pioneer Charles Lindbergh and his wife, Anne Morrow Lindbergh. The night before the mission, Lindbergh shared stories of his groundbreaking 1927 solo flight across the Atlantic, marveling at the comparison between his journey and Apollo 8's monumental endeavor. He reflected on the fuel requirements for his flight—a mere 450 gallons—versus the Saturn V's staggering fuel consumption, which was over 1,000 times greater. The next morning, on December 21, 1968, Lindbergh and his wife watched from a nearby dune as the Saturn V, with Borman, Lovell, and Anders aboard, lifted off from Kennedy Space Center, bound for lunar orbit.

Apollo 8 lifted off from Kennedy Space Center at 12:51 UTC (7:51 AM EST) on December 21, 1968, propelled into space by the three stages of the Saturn V rocket. The first stage, the S-IC, separated and splashed down in the Atlantic Ocean at 30°12′N 74°7′W. The second stage, the S-II, fell into the ocean at 31°50′N 37°17′W. The spacecraft's third stage, the S-IVB, performed the critical TLI burn before detaching from the command and service module (CSM).

Once in Earth orbit, the crew, supported by Houston's Mission Control, spent 2 hours and 38 minutes conducting spacecraft checks to ensure all systems were functioning properly for the upcoming TLI maneuver. At 2 hours, 27 minutes, and 22 seconds into the mission, the Capsule Communicator (CAPCOM) Michael Collins transmitted the message that would set Apollo 8 on its course to the Moon: "Apollo 8, you are go for TLI." The S-IVB engine ignited on schedule, flawlessly executing the burn. In five minutes, Apollo 8's velocity increased from 7,600 to 10,800 meters per second (25,000 to 35,000 feet per second).

Apollo 8 S-IVB rocket stage shortly after separation

Following the TLI burn, the Apollo 8 command and service modules separated from the spent S-IVB stage. The crew then rotated the spacecraft, taking photographs of the detached stage and conducting formation flying exercises with it. This marked the first time humans had witnessed the Earth in its entirety, visible in the distance as they traveled further away from their home planet.

As the spacecraft continued its journey, Apollo 8 Commander Frank Borman grew concerned that the S-IVB was staying too close to the CSM, prompting discussions with Mission Control about performing a separation maneuver. Ultimately, the crew conducted a burn using the service module's reaction control system (RCS), increasing their velocity by 7.7 feet per second (2.3 meters per second) in the Earthward direction, safely distancing themselves from the S-IVB. This maneuver, however, delayed the crew's onboard tasks by about an hour.

Five hours after launch, Mission Control sent a command to the S-IVB to vent its remaining fuel, ensuring it would pose no further risk to Apollo 8. The stage, along with a test article simulating the lunar module's weight and balance, bypassed the Moon and entered a solar orbit, where it remains to this day, orbiting the Sun on a 340.80-day cycle.

As the first crewed spacecraft to travel beyond low Earth orbit, Apollo 8 was also the first to pass through the Van Allen radiation belts. These belts, located up to 15,000 miles (24,000 kilometers) from Earth, contain trapped particles from solar wind that present a potential radiation hazard to astronauts. NASA scientists predicted that the fast transit through the belts would expose the crew to minimal radiation, comparable to a chest X-ray, with a dose of about 1 milligray (mGy).

To monitor radiation exposure, each astronaut wore a Personal Radiation Dosimeter, which transmitted data back to Earth, along with three passive film dosimeters that recorded cumulative exposure. By the mission's end, the crew had received an average radiation dose of 1.6 mGy, well within the expected safe range.

Jim Lovell's main responsibility as Command Module Pilot aboard Apollo 8 was navigating the spacecraft on its trajectory toward the Moon. While Mission Control handled most of the navigation calculations, a crew member needed to be skilled in celestial navigation in case of communication failure. Using a sextant built into the spacecraft, Lovell measured the angles between stars and the Earth's or the Moon's horizon. However, his task was complicated by debris surrounding the spacecraft, making distinguishing stars from particles of light reflecting off the debris cloud difficult.

About seven hours into the mission, the crew found themselves 1 hour and 40 minutes behind their flight plan. This delay resulted from difficulties in separating from the S-IVB stage and the complications Lovell faced with obscured star sightings. To ensure proper thermal regulation of the spacecraft, the crew initiated the Passive Thermal Control (PTC), commonly called the "barbecue roll." In this mode, the spacecraft rotated about once per hour around its long axis to evenly distribute heat. Without such control, the extreme temperatures encountered in space—up to 200°C (392°F) in sunlight and -100°C (-148°F) in shadow—could have caused the heat shield to crack or fuel lines to burst. Despite this system, minor adjustments were necessary every half-hour as the spacecraft gradually veered from its desired roll pattern.

Apollo 8 was a groundbreaking mission in the Apollo program, marking the first crewed spacecraft to leave Earth's orbit and journey to another celestial body—the Moon. Launched on December 21, 1968, it was a pivotal moment in the Space Race, demonstrating the United States' ability to achieve complex space travel objectives and paving the way for future lunar landings.

The mission began with the spacecraft entering a circular Earth parking orbit at an altitude of approximately 100 nautical miles (185 kilometers). The spacecraft's apogee was measured at 99.99 nautical miles (185.18 kilometers), and its perigee at 99.57 nautical miles (184.40 kilometers), with an orbital inclination of 32.51 degrees relative to the equator. The crew remained in this parking orbit for 2 hours and 44 minutes, conducting detailed system checks. During this time, a small amount of propellant vented, causing a slight increase in the spacecraft's altitude.

The translunar injection (TLI) maneuver marked the next crucial phase. Executed by the Saturn V's S-IVB third stage, the TLI burn lasted for 318 seconds, accelerating the spacecraft to 35,505 feet per second (10,822 meters per second)—a velocity just shy of Earth's escape velocity. This maneuver propelled Apollo 8 onto a translunar trajectory, making it the fastest human-crewed spacecraft at the time, a historic achievement in the quest to reach the Moon.

About 11 hours into the mission, the crew performed the first mid-course correction using the service propulsion system (SPS) engine. Though a helium bubble in the oxidizer lines caused a slight shortfall in the velocity change, the crew quickly compensated by using the reaction control system (RCS) thrusters. This fine-tuning ensured Apollo 8 remained perfectly on course for its lunar destination, eliminating the need for further mid-course corrections.

As mission commander Frank Borman took the first sleep shift, he encountered challenges in resting. The constant noise from radio communications and spacecraft systems made sleep difficult, prompting him to take a Seconal sleeping pill. Despite this, he awoke feeling nauseous and experienced vomiting and diarrhea, a situation complicated by the zero-gravity environment, where small particles of vomit and feces floated inside the spacecraft. Although initially reluctant to report his condition, Borman's crewmates, Jim Lovell and Bill Anders, encouraged him to inform Mission Control. Using the spacecraft's Data Storage Equipment (DSE), the crew recorded a message for the medical team on Earth. After reviewing the symptoms, doctors concluded Borman's illness was likely caused by either a brief viral infection or a reaction to the sleeping pill. Today, researchers suggest he may have been suffering from space adaptation syndrome, a common condition during the first few days in space as the body adjusts to weightlessness.

Despite this episode, the cruise phase of the mission proceeded smoothly. The crew routinely checked the spacecraft's systems and maintained their trajectory toward the Moon. One of the mission's most memorable moments came 31 hours after launch when the crew made the first television broadcast from space. Using a 2-kilogram black-and-white camera with a Vidicon tube, they attempted to give viewers on Earth a tour of the spacecraft. Although capturing clear images of Earth from space proved difficult, the broadcast captivated the public and lasted 17 minutes. Jim Lovell concluded the transmission by wishing his mother a happy birthday, adding a personal touch to this historic broadcast from deep space.

By the time of the first broadcast, the crew had abandoned their planned sleep shifts. Lovell slept 32.5 hours into the flight, ahead of schedule, while Anders followed shortly after taking a sleeping pill. Throughout much of the outward cruise, the crew was unable to see the Moon. The spacecraft's orientation for thermal control and fogging of the windows from out-gassed oils made it difficult to view the lunar surface. It wasn't until Apollo 8 traveled behind the Moon that the crew finally saw it.

A second television broadcast took place 55 hours into the mission. During this broadcast, the crew successfully captured and broadcast the first television images of Earth from space using filters from the still cameras. Though they had to maneuver the spacecraft to aim the camera, the crew described the Earth's vivid colors and features to viewers. This historic broadcast lasted 23 minutes, providing humanity with its first live televised glimpse of the Earth as seen from the vast expanse of space.

The challenges and accomplishments of Apollo 8's lunar trajectory phase underscored the complexity of navigating deep space while also showcasing the beauty of Earth from a perspective never before seen by humankind. These early broadcasts, coupled with the technical precision of the spacecraft's journey, laid the groundwork for future lunar exploration.

At 55 hours and 40 minutes into the Apollo 8 mission, the crew became the first humans to enter the gravitational sphere of influence of another celestial body, the Moon. This meant that the Moon's gravitational pull on the spacecraft was stronger than Earth's. At the time, Apollo 8 was 38,759 miles (62,377 km) from the lunar surface and traveled at 3,990 feet per second (1,220 m/s) relative to the Moon. Although this marked a historic milestone, the crew was more focused on maintaining their precise trajectory calculations, which were still referenced against their launch point at Kennedy Space Center. This continued until they performed the last mid-course correction, designed to position them for Lunar Orbit Insertion (LOI).

The final mid-course correction occurred before the spacecraft approached the Moon. This burn was retrograde, meaning it was against the direction of travel, and slowed the spacecraft by 2.0 feet per second (0.61 m/s). This small speed reduction adjusted the spacecraft's trajectory, ensuring it would pass just 71.7 miles (115.4 km) above the lunar surface. At precisely 61 hours after launch, while the spacecraft was approximately 24,200 miles (38,900 km) from the Moon, the reaction control system (RCS) thrusters fired for 11 seconds, completing this adjustment.

At 64 hours into the mission, the Apollo 8 crew began preparing for the critical Lunar Orbit Insertion 1 (LOI-1). This maneuver would slow the spacecraft sufficiently to allow it to be captured by the Moon's gravity and enter orbit. Due to the constraints of orbital mechanics, LOI-1 had to be performed on the far side of the Moon, where the crew would be out of radio contact with Earth. Before losing communication, Mission Control conducted a final "go/no go" poll to assess whether the mission was ready for this key event. At 68 hours into the flight, the crew received the green light from Mission Control with the message, "You are riding the best bird we can find." In a historic moment, Lovell responded, "We'll see you on the other side." For the first time in human history, astronauts ventured behind the Moon and out of radio contact with Earth, entering a period of tense silence for those following the mission.

Frances "Poppy" Northcutt, the first woman to work in NASA's Mission Control, recalled the nerve-wracking uncertainty during this time. Northcutt, who had helped calculate the mission's return trajectory, described the anxiety felt by the team when Apollo 8 went behind the Moon: "That was a very nerve-racking period... because of this thing with losing signal. You've got this big mystery going on there on the backside of the Moon. You do not know what's happening and there's not a darn thing anybody here can do about it until we hear from them."

With ten minutes remaining before the LOI-1 burn, the Apollo 8 crew completed a final check of their spacecraft systems, ensuring every switch was properly positioned. At this point, they caught their first glimpse of the Moon's surface. Having been traveling over the unlit side, Lovell noticed the first rays of sunlight striking the lunar landscape. However, with the LOI burn only two minutes away, there was little time to appreciate the view fully. The crew's attention swiftly turned to the critical task ahead: executing the burn that would place them in lunar orbit, setting the stage for the next chapter of human space exploration.

At 69 hours, 8 minutes, and 16 seconds after launch, the Apollo 8 crew ignited the service propulsion system (SPS) engine for Lunar Orbit Insertion (LOI-1). This critical burn, lasting 4 minutes and 7 seconds, placed the spacecraft in orbit around the Moon. The crew later described the experience as the longest four minutes of their lives, knowing that even a slight miscalculation could have catastrophic consequences. Had the burn been too short, the spacecraft would have entered a highly elliptical orbit or been flung into space. If too long, they could have crashed into the lunar surface. Fortunately, the burn was executed perfectly, and Apollo 8 settled into a stable lunar orbit.

As Mission Control waited anxiously on Earth, the signal confirming Apollo 8's successful insertion into lunar orbit came at the exact calculated moment. The spacecraft was now in an orbit measuring 193.3 by 69.5 miles (311.1 by 111.8 km). This precise orbit allowed the crew to begin their 20-hour survey of the Moon.

After verifying that all systems were functioning properly, Jim Lovell gave the first detailed description of the lunar surface, which appeared to him as a vast, colorless expanse:

"The Moon is essentially grey, no color; looks like plaster of Paris or sort of a grayish beach sand. We can see quite a bit of detail. The Sea of Fertility doesn't stand out as well here as it does back on Earth. There's not as much contrast between that and the surrounding craters. The craters are all rounded off. There's quite a few of them, some of them are newer. Many of them look like—especially the round ones—look like hit by meteorites or projectiles of some sort. Langrenus is quite a huge crater; it has a central cone. The crater's walls are terraced, about six or seven different terraces on the way down."

Upon arrival at the Moon, Apollo 8 performed a lunar orbit insertion (LOI), entering an elliptical orbit with a perilune (closest point) of 60.0 nautical miles (111.1 km) and an apolune (farthest point) of 168.5 nautical miles (312.1 km). The mission profile called for the spacecraft's orbit to be circularized at 60.7 by 59.7 nautical miles (112.4 by 110.6 km), resulting in an orbital period of 128.7 minutes. However, the uneven gravitational field of the Moon, due to mass concentrations or "mascons," slightly perturbed the spacecraft's orbit. For the ten lunar orbits, which lasted approximately twenty hours, Apollo 8's orbital distance shifted to 63.6 by 58.6 nautical miles (117.8 by 108.5 km).

At its farthest point from Earth, Apollo 8 reached a maximum distance of 203,752 nautical miles (377,349 kilometers), a record for the greatest distance humans had ever traveled from their home planet.

As they orbited the Moon, the crew's mission included reconnaissance of potential landing sites for future Apollo missions. Mare Tranquillitatis, the planned landing site for Apollo 11 was particularly interesting. The lighting conditions were ideal for examining the terrain, and Bill Anders took full advantage, photographing numerous points of interest. Over the course of the mission, the crew captured more than 800 high-resolution still photographs and recorded 700 feet (210 meters) of 16 mm film, documenting the lunar surface in unprecedented detail.

Despite the wonder of their surroundings, the crew remained focused on the spacecraft's critical systems. Borman, ever cautious, repeatedly inquired about the health of the SPS engine, which would be needed for their return to Earth. He requested a "go/no go" decision from Mission Control before passing behind the Moon on each orbit, ensuring that they would be ready for any contingency.

During their second orbit, the crew set up equipment to broadcast live images of the lunar surface back to Earth. Anders narrated the view as they passed over the Moon's craters, sharing with the world a vision of the Moon few had ever imagined. At the conclusion of this orbit, they executed a second burn, LOI-2, lasting 11 seconds, to further refine and circularize their orbit to 70.0 by 71.3 miles (112.7 by 114.7 km), providing an optimal path for continued observations.

For the next several orbits, the crew continued to check spacecraft systems and observe the Moon, capturing images and data for future missions. Frank Borman took a moment to fulfill a personal commitment during their third pass. He had been scheduled to participate in a prayer service at St. Christopher's Episcopal Church in Seabrook, Texas. Although his flight prevented him from attending in person, fellow parishioner Rod Rose, an engineer at Mission Control, suggested that Borman read a prayer during the flight, which could be recorded and played during the service. As Apollo 8 orbited the Moon, Borman recited the prayer, bringing a spiritual dimension to the mission's already historic achievements.

For the next 20 hours, the crew of Apollo 8 orbited the Moon, meticulously studying the surface and preparing for their eventual return to Earth. Their observations, data, and photographs would prove invaluable for future Apollo missions, including the landmark landing of Apollo 11 the following year.

On Apollo 8's fourth pass around the Moon, the crew witnessed a historic sight—the first "Earthrise" observed by human eyes. As the spacecraft emerged from behind the lunar surface, Bill Anders saw the Earth rising above the lunar horizon and excitedly called to his fellow astronauts. He quickly captured the moment in a black-and-white photograph before requesting color film to take what would become the iconic Earthrise image. This photo, showing the blue and white Earth suspended against the vastness of space and the barren grey lunar surface, would later be chosen by Life magazine as one of the 100 most important photographs of the century. While NASA's Lunar Orbiter 1 had previously taken a photo of Earthrise in 1966, this was the first time humans experienced it firsthand.

Earthrise, a phenomenon made possible by the synchronous rotation of the Moon, is not visible from any fixed point on the lunar surface. From a single spot, Earth remains stationary in the sky due to the Moon's rotation matching its orbit around Earth. Only astronauts in orbit around the Moon or near specific points on the lunar limb can witness this stunning event, where the Earth appears to rise and fall above the lunar horizon due to slight variations in the Moon's position, called libration.

As Anders continued photographing the lunar surface, Lovell took control of the spacecraft, allowing Frank Borman to rest. Despite the noise and cramped conditions aboard the spacecraft, Borman managed to sleep for two lunar orbits. However, he was soon awakened by his crew's fatigue-induced mistakes. Realizing that none of them had slept properly in over three days, Borman ordered Lovell and Anders to rest. Though Anders initially protested, he eventually agreed, requesting that Borman set the camera to automatically take pictures while he slept. Borman complied, as he knew the importance of staying alert for the upcoming second television broadcast.

During their ninth orbit, the astronauts began their second broadcast, transmitting to a global audience on Christmas Eve, 1968. Borman introduced the crew, and each member shared their thoughts about the lunar surface and their experience orbiting the Moon. Borman described the landscape as a "vast, lonely, forbidding expanse of nothing." As the astronauts provided their observations, Anders informed viewers that they had a special message to share. Each astronaut then read from the Book of Genesis, reciting the story of creation. Borman concluded the broadcast with a heartfelt Christmas message to the world: "And from the crew of Apollo 8, we close with good night, good luck, a Merry Christmas, and God bless all of you—all of you on the good Earth." This broadcast, deeply moving in its simplicity and significance, became one of the most memorable moments of the Apollo 8 mission.

the crew taken while they were in orbit around the Moon.
Frank Borman is in the center.

With the broadcast complete, the crew prepared for their final critical task in lunar orbit—the trans-Earth injection (TEI), scheduled to occur about two and a half hours after the end of the transmission. This maneuver would propel them out of lunar orbit and set them on a course back to Earth. As with their earlier burns, TEI had to be performed on the far side of the Moon, where they were out of radio contact with Mission Control. The SPS engine ignited exactly on time, and telemetry was reacquired when the spacecraft reappeared from behind the Moon at 89 hours, 28 minutes, and 39 seconds—precisely as calculated. Upon regaining voice contact, Jim Lovell humorously reported, "Please be informed, there is a Santa Claus," to which CAPCOM Ken Mattingly replied, "That's affirmative, you are the best ones to know."

Having successfully completed the burn, Apollo 8 began its journey back to Earth on Christmas Day, December 25. The mission, which had made humanity's first voyage to another celestial body, now turned toward home, having forever changed the perspective from which humans viewed their planet and the cosmos.

The first image taken by humans of the whole Earth disk,

In the midst of their mission, Jim Lovell, taking advantage of some free time, decided to conduct navigational sightings by aligning the Apollo 8 spacecraft with various stars. Using the computer's onboard keyboard, Lovell inadvertently erased some critical memory from the guidance system. This action caused the inertial measurement unit (IMU) to assume the spacecraft was still in its pre-launch orientation, leading the thrusters to adjust the spacecraft's attitude in an attempt to "correct" the perceived error.

Once the crew recognized what had happened, they realized they would need to input the correct orientation data into the computer manually. Lovell spent ten minutes calculating the necessary figures, aligning the spacecraft with the stars Rigel and Sirius, and an additional fifteen minutes entering the new data into the computer. This experience would serve as valuable preparation for Lovell, who would face a similar but far more critical manual realignment during the Apollo 13 mission, following an IMU shutdown to conserve power.

After their successful mission orbiting the Moon, the Apollo 8 crew began the return journey to Earth. The next two-and-a-half days were relatively calm, allowing the astronauts to relax and monitor the spacecraft's systems as they cruised toward reentry. If the trajectory specialists' calculations were accurate, Apollo 8 would enter Earth's atmosphere on schedule and splash down in the Pacific Ocean.

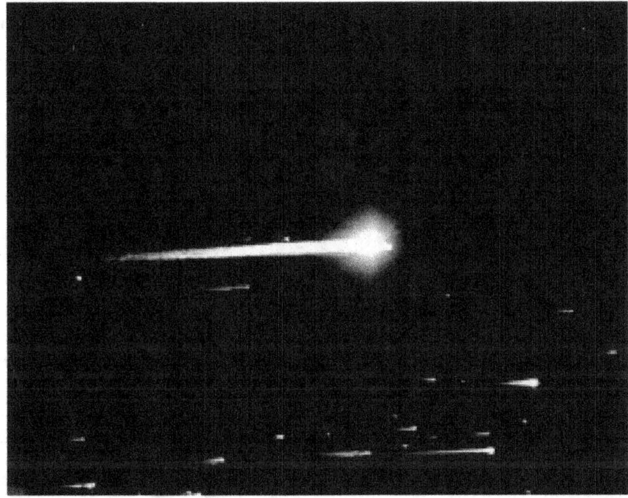

On Christmas Day, the crew conducted their fifth televised broadcast, providing viewers back on Earth with a glimpse of life aboard a spacecraft. They demonstrated how astronauts lived and worked in space, and shared a holiday surprise—a turkey dinner, complete with stuffing, courtesy of Deke Slayton, which had been packed especially for the occasion. Another surprise came in the form of three miniature bottles of brandy, though Commander Frank Borman instructed the crew to save them until after the mission was complete.

The following day, Apollo 8's sixth and final television broadcast featured breathtaking images of Earth, offering a stunning view of their home planet as they neared the end of their mission. With the computer set to control the reentry, the crew's role was simply to ensure the spacecraft was properly aligned for the descent.

As the command module separated from the service module, the heat shield was exposed, readying the spacecraft for the intense heat of reentry. The service module, no longer needed, was left to burn up in the atmosphere. Six minutes before reentry, the astronauts saw the Moon rise above Earth's horizon—a symbolic moment marking the completion of their historic mission.

As the spacecraft entered the thin outer atmosphere, plasma formed around the module, casting a glowing haze outside the windows. The Apollo 8 crew felt the force of deceleration increase as the spacecraft slowed down, reaching a peak of 6 g (59 m/s^2). With the computer controlling the descent, the spacecraft briefly rose like a skipping stone before continuing its descent toward the Pacific Ocean. At an altitude of 30,000 feet, the drogue parachutes deployed, followed by the three main parachutes at 10,000 feet, ensuring a stable descent into the ocean. Apollo 8 splashed down at 15:51 UTC on December 27, 1968, at 8°8′N, 165°1′W, in the North Pacific Ocean, southwest of Hawaii.

Upon splashdown, the command module landed upside down, a position known as "Stable 2." As the astronauts were tossed by 10-foot swells, Frank Borman became nauseous. However, the flotation bags automatically deployed, righting the spacecraft to the "Stable 1" position within six minutes. Rescue operations were swift, with the first recovery swimmer arriving 43 minutes after splashdown. Within another 45 minutes, the crew was safely aboard the USS Yorktown, where they were greeted with cheers and relief after their historic mission.

Apollo 8's mission came at the end of a turbulent 1968, a year marked by political assassinations, civil unrest, and international conflicts. In the face of such upheaval, the successful lunar orbit mission offered a glimmer of hope and unity. Time magazine named the Apollo 8 astronauts—Frank Borman, Jim Lovell, and Bill Anders—its Men of the Year for 1968, acknowledging their achievement as one that profoundly influenced the course of human history.

The crew of Apollo 8 were the first humans to leave Earth's gravitational influence and orbit another celestial body. Their mission, which many—including the astronauts themselves—rated as having only a fifty-fifty chance of success, had been a daring and groundbreaking endeavor. After the mission, a telegram sent to Borman encapsulated the mission's profound impact on the world: "Thank you, Apollo 8. You saved 1968."

One of the most enduring legacies of the Apollo 8 mission was the iconic "Earthrise" photograph, captured as the spacecraft emerged from behind the Moon during its fourth orbit. This image, showing the Earth as a fragile blue sphere rising above the barren lunar surface, would later become a symbol of environmental consciousness, inspiring the first Earth Day in 1970. Life magazine selected it as one of its "100 Photographs That Changed the World."

Apollo 8 was widely covered by global media, with more than 1,200 journalists reporting on the mission. The mission's Christmas Eve broadcast from lunar orbit reached an estimated one-quarter of the world's population. Soviet leader Boris Petrov praised the mission, calling it an "outstanding achievement of American space sciences and technology." Apollo 8's broadcasts earned the crew an Emmy Award, recognizing their contribution to space exploration and public engagement.

The mission also sparked controversy when atheist activist Madalyn Murray O'Hair filed a lawsuit against NASA, objecting to the crew's public reading of the Book of Genesis during the Christmas Eve broadcast. Although the lawsuit was eventually dismissed, it led to NASA's cautious approach to religious expression in future space missions. Buzz Aldrin, during Apollo 11, quietly conducted a private Communion ceremony on the Moon, choosing not to publicize the event at the time.

In recognition of the mission's historical significance, the U.S. Postal Service issued a commemorative stamp in 1969 featuring the Earthrise photograph and the words "In the beginning God..." from the Genesis reading. Apollo 8's legacy is also reflected in popular culture, as the crew participated in the pre-game show of Super Bowl III shortly after their return, further cementing their place in the collective memory of a transformative period in space exploration.

Apollo 8 marked a turning point in the Space Race, demonstrating that humans could not only reach the Moon but also safely return. Just months later, this mission paved the way for the Apollo 11 landing, which remains one of the most celebrated achievements of the Apollo program.

Apollo 8's mission forever changed the course of human space exploration. It demonstrated that reaching the Moon was within grasp and laid the groundwork for the historic lunar landing that followed less than a year later. The three astronauts, Borman, Lovell, and Anders, became national heroes, and their journey into lunar orbit captured the imagination of people around the globe.

As of June 2024, Jim Lovell is the sole surviving member of the Apollo 8 crew, following the deaths of Frank Borman in November 2023 and William Anders in June 2024. Their legacy, however, lives on as a testament to the daring and determination that defined NASA's Apollo program and America's ambition during the Space Race.

Venera 5

First parachute to be deployed on another planet (Venus)

Venera 5, part of the Soviet Union's ambitious Venera program, was designed to explore and gather data from Venus, Earth's closest planetary neighbor. Launched on January 5, 1969, from the Baikonur Cosmodrome aboard a Molniya-M rocket, Venera 5 was a critical mission in the Soviet space program aimed at expanding humanity's understanding of the Venusian atmosphere.

The spacecraft was a robust, upgraded version of its predecessor, Venera 4, specifically built to withstand the harsh conditions of Venus' atmosphere. The mission's primary goal was to obtain precise atmospheric data, a task that Venera 4 had started but could not complete due to the extreme pressures and temperatures on the planet. Venera 5 was equipped with advanced scientific instruments to continue this work, providing more accurate chemical analysis of the atmosphere.

Upon nearing Venus on May 16, 1969, a descent capsule weighing 405 kilograms (893 pounds) was jettisoned from the main spacecraft. This capsule contained a suite of scientific instruments designed to study the Venusian environment during its descent. A parachute deployed to slow the capsule's descent, allowing it to transmit data for 53 minutes before the crushing atmospheric pressure and intense heat rendered it inoperative.

The spacecraft's descent through the dense Venusian atmosphere was measured and analyzed by several instruments. The KS-18-3M instrument studied cosmic particle flows, while the LA-2U instrument analyzed the distribution of oxygen and hydrogen in the atmosphere. As the lander approached Venus' surface, instruments like the G-8 gas analyzers provided critical data on the atmosphere's chemical composition, confirming the presence of high concentrations of carbon dioxide. Pressure sensors, such as the MDDA A, recorded atmospheric pressures ranging from 13 to 4,000 kilopascals, while temperature measurements from the CE-164D instrument documented readings as high as 320°C (608°F).

The parachute system, a key part of the capsule's design, had been modified based on lessons learned from Venera 4. Engineers designed smaller parachutes for Venera 5, ensuring that the capsule descended more quickly, allowing it to reach greater depths of Venus' atmosphere before its instruments were overwhelmed by the extreme conditions.

Venera 5 entered Venus' nightside atmosphere at an altitude of 37,000 kilometers (23,000 miles) and at a speed of 210 meters per second (472 mph). The parachute deployed at this point, slowing the capsule and initiating the data transmission sequence. For 53 minutes, the probe sent vital atmospheric data back to Earth in 45-second intervals, providing a clearer understanding of Venus' hostile environment. These transmissions confirmed the high pressures and temperatures recorded by Venera 4, as well as the overwhelming presence of carbon dioxide in the planet's atmosphere.

The spacecraft also carried symbolic items to Venus: a medallion bearing the State Coat of Arms of the Soviet Union and a bas-relief of Lenin, which were delivered to the night side of the planet, marking a significant moment in Soviet space exploration.

Instruments aboard Venera 5 confirmed that Venus was enveloped in a thick, hostile atmosphere with surface conditions too extreme for human survival. These findings were a major milestone in planetary science, contributing valuable insights to the understanding of Venus, and establishing the Venera program as a crucial chapter in the early years of space exploration.

Soyuz 4 /Soyuz 5

First crew exchange in space
First docking of two manned spacecraft

Soyuz 4, launched on January 14, 1969, marked a significant moment in the Soviet space program. Commanded by Vladimir Shatalov on his first flight, the mission aimed to dock with Soyuz 5 and transfer two crew members, Aleksei Yeliseyev and Yevgeny Khrunov, from Soyuz 5 back to Soyuz 4 for the return to Earth. This complex maneuver had been attempted unsuccessfully during the Soyuz 3 mission, making Soyuz 4 and 5's mission critical for advancing Soviet space capabilities.

The radio call signs for the two spacecraft were "Amur" (Soyuz 4) and "Baikal" (Soyuz 5), symbolizing the Baikal-Amur Mainline railway project, a major infrastructure initiative in the Soviet Union at the time. Shatalov, aboard Soyuz 4, was tasked with orchestrating the docking, while Yeliseyev and Khrunov would undertake a spacewalk (extravehicular activity or EVA) to transfer between the two spacecraft—a requirement as the Soyuz design had no internal docking tunnel.

Following the docking failure of Soyuz 2 and Soyuz 3 in late 1968, Soviet engineers revised docking procedures for the Soyuz 4 and 5 missions. These adjustments included ensuring the docking would occur in daylight and over Soviet territory, where ground communication was optimal. Despite the challenging conditions of mid-winter Kazakhstan, with temperatures plummeting to -25°C (-13°F), preparations continued at the Baikonur Cosmodrome. Amid freezing temperatures, workers dealt with illnesses such as influenza and frostbite, and even a few tragic deaths among military personnel.

Soyuz 4 and Soyuz 5 after performing the first docking
of two crewed spacecraft on 16 January 1969

After the successful launch of the Venera 6 mission on January 10, attention shifted to the Soyuz spacecraft. However, during the countdown for Soyuz 4, a malfunction in the booster's roll rate gyroscope was detected, causing a temporary delay. Engineers resolved the issue, likely caused by ice inhibiting electrical contact, and Soyuz 4 launched successfully on January 14, 1969, at 10:30 AM local time. The spacecraft reached orbit with all systems functioning as planned.

On January 16, 1969, Soyuz 4 and Soyuz 5 made history by achieving the first successful docking between two crewed spacecraft, a critical step toward future space station missions. The docking mechanism was primitive by modern standards, consisting of a probe and drogue system with no tunnel for internal transfer. This required Yeliseyev and Khrunov to perform an EVA, spacewalking from Soyuz 5 to Soyuz 4.

Yeliseyev and Khrunov, donned in their Yastreb spacesuits, began preparing for the EVA shortly after the docking. The Yastreb suit, designed to prevent ballooning in the vacuum of space, featured a cable-and-pulley system to aid mobility and a regenerative life support system worn on the chest. The design was finalized after the difficulties encountered by Alexei Leonov during the first-ever spacewalk in 1965.

The EVA commenced on January 16 at 12:43 GMT. Khrunov was the first to exit Soyuz 5, moving across the void of space to Soyuz 4, followed shortly by Yeliseyev. Although the EVA was a success, one of Khrunov's lines became tangled during the spacewalk, momentarily causing concern. Additionally, due to the distraction, Yeliseyev failed to set up the onboard camera, leaving only poor-quality video transmissions of this historic moment. Despite these setbacks, the two cosmonauts safely entered Soyuz 4, where Shatalov greeted them after repressurizing the spacecraft's orbital module, which served as an airlock.

The two Soyuz spacecraft remained docked for 4 hours and 35 minutes before separating. Soyuz 4, now carrying Shatalov, Yeliseyev, and Khrunov, successfully re-entered Earth's atmosphere and landed safely on January 17, 1969, near Karaganda, Kazakhstan. Soyuz 5, with Boris Volynov aboard, followed suit, though its re-entry was far more perilous due to a malfunction, with Volynov narrowly escaping disaster.

The success of Soyuz 4 and 5 demonstrated the feasibility of performing complex docking and crew transfer maneuvers, both essential elements for a Soviet lunar landing. The mission rehearsed procedures that would have been used for a manned lunar mission, in which a cosmonaut would have landed on the Moon and then returned to orbit, requiring a spacewalk between the lunar lander and the orbiting spacecraft, as the Soviet spacecraft lacked an internal transfer tunnel like the Apollo Lunar Module.

Despite the mission's triumph, the Soyuz 4/5 crew's planned meeting with Soviet Premier Leonid Brezhnev was marred by an attempted assassination. On the day of their celebratory reception at the Kremlin, a man fired several shots at Brezhnev's motorcade, though the target was mistakenly believed to be a group of famous cosmonauts, including Alexei Leonov and Valentina Tereshkova. The Soyuz 4 and 5 crews, waiting to greet Brezhnev, witnessed the motorcade's hasty departure after the incident.

The Soyuz 4 and 5 missions stand as a testament to Soviet ingenuity during the Cold War space race, as they accomplished several historic firsts, including the first docking of two crewed spacecraft and the second Soviet spacewalk. The lessons learned from this mission directly influenced the development of future Soviet space endeavors, including space station operations and lunar exploration strategies, despite the eventual shift in focus following the U.S. success with Apollo 11. These missions laid the groundwork for the long-term use of the Soyuz spacecraft in Soviet and, later, Russian space programs, continuing into the era of the International Space Station.

Apollo 11

First humans on the Moon
First space launch from another celestial body
First sample return from the Moon

On February 8, 1968, NASA's Apollo Site Selection Board announced five potential landing sites for the upcoming lunar mission. These locations were the culmination of two years of intensive studies based on high-resolution photography captured by the five uncrewed probes of the Lunar Orbiter program. These photographs, combined with surface condition data from the Surveyor program, helped overcome the limitations of Earth-based telescopes, which could not provide the resolution required for Apollo's precise landing needs.

The chosen landing site had to meet several criteria: proximity to the lunar equator to minimize propellant use, a relatively obstacle-free surface to reduce maneuvering, and a flat terrain to simplify the task of the landing radar. Scientific value was not prioritized in these early considerations. Promising areas on Earth-based photographs were often deemed unsuitable upon closer inspection, leading to the relaxation of the initial requirement that the site be completely free of craters, as no such location could be found.

Five candidate sites emerged: Sites 1 and 2 were in the Sea of Tranquility (Mare Tranquillitatis), Site 3 was in the Central Bay (Sinus Medii), and Sites 4 and 5 were in the Ocean of Storms (Oceanus Procellarum). The final selection was based on seven key criteria, including a smooth surface with relatively few craters, approach paths free from large hills or cliffs, a minimal propellant requirement, and good visibility during the landing approach. Additionally, the Sun's position had to be between 7 and 20 degrees behind the Lunar Module (LM), and the general slope of the landing area could not exceed two degrees. These requirements made the landing window particularly restrictive, limiting the mission to just one day per month.

Site 2 in the Sea of Tranquility was ultimately chosen, with Sites 3 and 5 as backups. In May 1969, the Apollo 10 mission's Lunar Module flew to within 15 kilometers (9.3 miles) of Site 2, confirming its suitability for Apollo 11's historic landing.

As Apollo 11 preparations continued, one question loomed large: Who would be the first astronaut to step onto the lunar surface? The crew's first press conference directly addressed this question, but at the time, both Commander Neil Armstrong and Lunar Module Pilot Buzz Aldrin indicated the decision had not been made and was not based on personal preference.

Initial plans had the lunar module pilot exit the spacecraft first, as had been the protocol during previous Gemini missions. Reporters speculated that Aldrin would be the first, and even NASA Associate Administrator George Mueller suggested as much. However, internally, there was debate. Aldrin lobbied hard for the honor, believing his status as a career astronaut made him the better choice, while rumors circulated that Armstrong, as a civilian astronaut, would be selected to make the first step.

Ultimately, NASA decided that Armstrong, as the mission commander, would exit first. This decision was made partly due to the physical layout of the Lunar Module. The hatch's location made it difficult for the astronauts, encumbered by bulky spacesuits, to maneuver. Armstrong, positioned closer to the exit, could more easily leave the module. A simulation where Aldrin attempted to exit first resulted in damage to the module, reinforcing the decision. Despite this, the official decision was not communicated to Armstrong and Aldrin until the spring of 1969, when Armstrong accepted the role, stating, "Yes, that's the way to do it."

On April 14, 1969, NASA publicly confirmed Armstrong would be the first to walk on the Moon, a decision that sparked controversy and media speculation. Some accused Armstrong of using his commander's prerogative to claim the honor. However, mission leaders, including Chris Kraft, later revealed that the decision was made to ensure that the first person to walk on the Moon would be someone with the calm and composed demeanor of Charles Lindbergh, traits Armstrong personified.

As the July 1969 launch date approached, preparations intensified. The ascent stage of Lunar Module 5 (Eagle) arrived at Kennedy Space Center on January 8, followed by the descent stage and Command and Service Module (CSM) Columbia later that month. Several improvements were made to the Lunar Module based on lessons learned from previous missions. Eagle was equipped with a VHF radio for communications during the lunar surface EVA, a lighter ascent engine, and enhanced thermal protection on the landing gear. The Early Apollo Scientific Experiments Package (EASEP) package of scientific experiments was also added to the mission's objectives.

By April 14, the Command and Service Modules had been integrated, and on May 20, the massive Saturn V rocket, with Apollo 11 aboard, began its slow journey from the Vehicle Assembly Building to Launch Pad 39A. Weighing over 5,400 tonnes, the fully assembled rocket was a marvel of engineering, dwarfing everything around it. The final countdown tests began on June 26, 1969, and concluded on July 2. By the evening of July 15, the crawler-transporter had carried the mobile service structure away from the launch pad, and the Saturn V stood ready for its historic mission.

NASA's Apollo 11 mission, which culminated in humanity's first landing on the Moon, involved a dedicated and skilled team of astronauts and support personnel, each playing a critical role in this historic achievement.

The prime crew of Apollo 11, officially announced on November 20, 1967, included Commander Neil Armstrong, Command Module Pilot (CMP) Michael Collins, and Lunar Module Pilot (LMP) Edwin "Buzz" Aldrin Jr. This was the second and final spaceflight for all three astronauts.

Armstrong, Aldrin, and Collins had diverse experiences in spaceflight. Armstrong, the first civilian astronaut in the mission, had flown in Gemini 8, where he made history by successfully docking two spacecraft. Aldrin, a career astronaut, had flown alongside Jim Lovell during Gemini 12, conducting extravehicular activities that set the stage for future Moon landings. Collins had initially been slated to fly on Apollo 8, but after developing a bony growth between his fifth and sixth vertebrae, he underwent surgery, and Lovell took his place on that mission. Once recovered, Collins rejoined the Apollo 11 crew, replacing Lovell as the CMP.

The Apollo 11 lunar landing mission crew, pictured from left to right, Neil A. Armstrong, commander; Michael Collins, command module pilot; and Edwin E. Aldrin Jr., lunar module pilot.

Despite the individual qualifications of the crew members, their working relationship was more professional than personal. Armstrong was known for his quiet, reserved demeanor, while Collins considered himself a loner, and Aldrin's attempts to create a closer bond with his crewmates did not take hold. Armstrong, however, expressed confidence in their collective abilities, emphasizing that all the crews he worked with, including Apollo 11, operated effectively as a team.

The backup crew for Apollo 11 consisted of James Lovell as Commander, William Anders as CMP, and Fred Haise as LMP. Lovell and Anders had flown together on the groundbreaking Apollo 8 mission, the first crewed spacecraft to orbit the Moon. As 1969 progressed, Anders accepted a position with the National Aeronautics and Space Council and announced his retirement as an astronaut, effective August 1969. To account for the possibility of Apollo 11 being delayed, Ken Mattingly trained alongside Anders as a potential backup CMP.

In the regular crew rotation system established for Apollo, Lovell, Mattingly, and Haise were originally slated for Apollo 14. However, due to various scheduling changes, they were reassigned to Apollo 13, where Lovell would lead the mission that became famous for its in-flight emergency.

For Apollo missions, a third group of astronauts was assigned to each mission—the support crew. This group was responsible for maintaining flight plans, checklists, and mission ground rules. They were tasked with developing emergency procedures, ensuring that both the prime and backup crews could focus on practicing and mastering them in simulations. The Apollo 11 support crew included Ken Mattingly, Ronald Evans, and Bill Pogue, all of whom later flew on other Apollo missions.

Capsule Communicators (CAPCOM)

The capsule communicator (CAPCOM) was a critical role during the Apollo missions, as this individual was the only person who communicated directly with the astronauts in space. For Apollo 11, several astronauts served as CAPCOMs, including Charles Duke, Ronald Evans, Bruce McCandless II, James Lovell, William Anders, Ken Mattingly, Fred Haise, Don L. Lind, Owen K. Garriott, and Harrison Schmitt. CAPCOMs played a vital role, particularly during critical phases of the mission, such as the lunar landing and extravehicular activities.

Apollo 11's flight directors were responsible for overseeing the various phases of the mission from NASA's Mission Control Center. Clifford E. Charlesworth led the team during the launch and the extravehicular activity (EVA) on the lunar surface, with Gerald D. Griffin serving as his backup. Gene Kranz, known for his calm leadership style, directed the lunar landing itself, while Glynn Lunney oversaw the lunar ascent. Milton Windler coordinated the planning aspects of the mission.

Numerous other individuals made significant contributions to the success of Apollo 11. Farouk El-Baz, a geologist, studied the Moon's surface and helped identify potential landing sites. Kurt Debus, a prominent rocket scientist, supervised the construction of launch infrastructure at Kennedy Space Center. Several individuals were responsible for critical aspects of the mission's logistics and planning: Jack Garman, a computer engineer, played a crucial role in troubleshooting technical issues, while Eldon C. Hall was instrumental in designing the Apollo Guidance Computer hardware. Margaret Hamilton led the team that developed the onboard flight software.

Gene Shoemaker, a geologist, trained the astronauts in field geology to prepare them for the lunar surface. Additionally, Bill Tindall coordinated the mission techniques that ensured smooth operations between teams. Millicent Goldschmidt, a microbiologist, developed aseptic lunar material collection techniques and trained astronauts in their use. Even the tailored spacesuits were a feat of engineering, designed by Eleanor Foraker, ensuring that the astronauts could safely explore the Moon's surface.

In the early stages of mission planning for Apollo 11, the crew initially adopted the whimsical names "Snowcone" for the Command Module (CM) and "Haystack" for the Lunar Module (LM) in both internal and external communications. However, following the Apollo 10 crew's decision to name their spacecraft "Charlie Brown" and "Snoopy," Julian Scheer, the Assistant Manager for Public Affairs, urged a more serious approach. He suggested the Apollo 11 crew choose names with deeper significance for such a historic mission.

Ultimately, the LM was named "Eagle" after the prominent motif in the mission's insignia, symbolizing American pride and ambition. The CM was named "Columbia," a name rooted in American history and exploration. According to astronaut Michael Collins, "Columbia" was chosen not only for its connection to the United States but also as a tribute to Jules Verne's 1865 novel From the Earth to the Moon, in which a spacecraft named Columbiad was launched from Florida. In addition, Collins mentioned in his 1976 book that the name also honored the legacy of Christopher Columbus, aligning the mission with the spirit of discovery and exploration.

Each Apollo astronaut was allowed to bring personal mementos in their Personal Preference Kits (PPKs), small bags containing items of personal significance. Five PPKs were carried on Apollo 11: three for the astronauts aboard the Command Module Columbia and two aboard the Lunar Module Eagle.

Neil Armstrong's PPK held artifacts of profound historical significance, including a piece of wood from the Wright brothers' 1903 Wright Flyer—the first powered aircraft—and a piece of fabric from its wing. These items symbolized the continuity of human flight, from the Wright brothers' groundbreaking achievement to Armstrong's journey to the Moon. Armstrong also carried a diamond-studded astronaut pin that had been intended for the ill-fated Apollo 1 mission. The widows of the Apollo 1 crew had gifted the pin to Deke Slayton, a senior NASA official, and Armstrong carried it with him to honor the memory of those astronauts.

Selecting a landing site for Apollo 11 was a meticulous process informed by years of study. On February 8, 1968, NASA's Apollo Site Selection Board announced five potential landing sites based on data from the Lunar Orbiter program, which provided high-resolution photographs of the Moon's surface, and the Surveyor program, which offered valuable information on surface conditions.

The landing site had to meet stringent criteria. It needed to be near the lunar equator to minimize fuel consumption, free of large obstacles, and relatively flat to simplify the landing process. The site had to be reachable with minimal fuel and have a slope of less than two degrees to ensure a stable landing. Additionally, the Sun's angle during landing was critical, requiring the Sun to be between 7 and 20 degrees behind the Lunar Module to provide optimal visibility.

NASA initially identified five potential sites, including two in the Sea of Tranquility (Mare Tranquillitatis), one in the Central Bay (Sinus Medii), and two in the Ocean of Storms (Oceanus Procellarum). Site 2, located in the Sea of Tranquility, was ultimately chosen due to its smooth surface and fewer large craters. Apollo 10 later flew a near-landing mission, passing within 15 kilometers of Site 2, confirming its suitability.

From the moment the Apollo 11 crew was introduced, one question dominated public curiosity: Who would be the first man to set foot on the Moon? At the time, NASA had not made an official decision. According to Deke Slayton, the crew assignments were not based on individual preferences, but on operational considerations.

Initially, the egress procedures from the Lunar Module (LM) followed the pattern established in the Gemini missions, where the Lunar Module Pilot (LMP) exited first. Early reports even suggested that Buzz Aldrin, the LMP, would be the first to walk on the Moon. However, tensions arose when it was rumored that Neil Armstrong, as Commander, would take the first step. Aldrin, upset by the decision, attempted to lobby for a change, but the decision had already been made.

The final decision to have Armstrong exit the Lunar Module first was influenced by both operational and symbolic factors. The cramped interior of the LM made it physically difficult for Aldrin to exit first, as Armstrong was positioned closer to the hatch. A simulation test where Aldrin attempted to leave first had caused damage to the spacecraft's interior, solidifying the decision.

NASA also believed Armstrong, with his calm demeanor and civilian status, was more fitting as the first to make the historic step. Comparisons were made to Charles Lindbergh, and mission planners wanted the first person on the Moon to embody the quiet, confident spirit of American exploration. On April 14, 1969, NASA officially announced that Armstrong would be the first to set foot on the lunar surface.

In later years, both Aldrin and Armstrong downplayed the controversy, with Armstrong emphasizing that the decision was based on logistical necessity rather than personal ambition. Nonetheless, the media speculated that Armstrong had used his position as Commander to ensure he was first. However, as revealed in memoirs such as Chris Kraft's autobiography, the decision had always been strategic, aiming to present the image of a calm, steady leader in the historic moment.

The preparation for Apollo 11 began early in 1969 with the arrival of key components at the Kennedy Space Center. On January 8, 1969, the ascent stage of Lunar Module (LM) 5, later named Eagle, arrived, followed by its descent stage on January 12. The Command and Service Module (CSM) 107, named Columbia, reached the center on January 23. These critical spacecraft components differed slightly from those used in previous missions. For instance, Eagle was outfitted with a VHF radio antenna to enable communication during the astronauts' extravehicular activities (EVAs) on the lunar surface. Additionally, it featured a lighter ascent engine, more thermal protection on the landing gear, and a scientific package known as the Early Apollo Scientific Experiments Package (EASEP). Columbia underwent minor modifications, with insulation removed from its forward hatch.

By January 29, the CSM was mated, and on April 14, it was moved from the Operations and Checkout Building to the Vehicle Assembly Building (VAB). Meanwhile, the massive Saturn V rocket that would carry Apollo 11 into space arrived in stages. The S-IVB third stage arrived on January 18, followed by the S-II second stage on February 6, the S-IC first stage on February 20, and the Saturn V Instrument Unit on February 27.

On May 20, the fully assembled Saturn V, weighing an immense 5,443 tonnes, rolled out of the VAB atop the crawler-transporter. It made its slow, deliberate journey to Launch Complex 39A, just as Apollo 10 was completing its mission to the Moon. A series of countdown tests began on June 26 and concluded on July 2. The launch complex was floodlit on the evening of July 15, as the final preparations were underway for the historic mission.

In the early hours of July 16, the S-II and S-IVB stages were filled with liquid hydrogen, completing fueling just three hours before launch. At 04:00, Deke Slayton, Director of Flight Crew Operations, woke the crew, who followed the traditional pre-flight ritual of showering and shaving before enjoying a hearty breakfast of steak and eggs with Slayton and the backup crew.

By 06:30, Neil Armstrong, Michael Collins, and Buzz Aldrin donned their space suits and began breathing pure oxygen. The crew arrived at Launch Complex 39A, where astronaut Fred Haise assisted them as they entered Columbia. Armstrong, Collins, and Aldrin took their positions on the left, right, and center couches, respectively. By 06:54, the hatch was sealed, and the cabin was pressurized. The countdown continued smoothly, transitioning to an automated process just over three minutes before launch. In the firing room, over 450 personnel monitored every aspect of the operation.

On the morning of July 16, 1969, the crew of Apollo 11, Neil Armstrong, Buzz Aldrin, and Michael Collins, awoke early to prepare for launch. After a traditional pre-flight breakfast of steak and eggs, the astronauts donned their space suits and began breathing pure oxygen. They made their way to Launch Complex 39, where they were seated in the cramped confines of the Command Module. Armstrong took his position on the left-hand couch, followed by Collins on the right, and Aldrin in the center.

As the final countdown commenced, the closeout crew sealed the hatch and pressurized the cabin. At three minutes and twenty seconds before launch, the countdown switched to automatic, with over 450 personnel monitoring the complex systems that would send the Apollo 11 crew on their historic journey to the Moon.

On July 16, 1969, at precisely 9:32 a.m. EDT (13:32:00 UTC), the world watched as the Saturn V rocket carrying Apollo 11 lifted off from Kennedy Space Center's Launch Complex 39A. An estimated one million spectators gathered near the site, filling highways and beaches to witness history. Among the dignitaries were General William Westmoreland, four U.S. Cabinet members, 19 state governors, 40 mayors, 60 ambassadors, 200 congressmen, and Vice President Spiro Agnew, who viewed the launch alongside former President Lyndon B.

Johnson and Lady Bird Johnson. Around 3,500 media representatives from 55 countries were present, and millions more watched live television broadcasts in 33 countries, with 25 million viewers in the United States alone.

The Saturn V rocket, designated AS-506, began its mission by rolling into a flight azimuth of 72.058° just over 13 seconds after launch. Two minutes and 42 seconds into the flight, the first-stage engines shut down, and the massive S-IC stage separated. The second stage, S-II, ignited and propelled the spacecraft into the upper atmosphere before it too shut down and separated approximately nine minutes into the mission. The S-IVB third stage then fired, placing Apollo 11 into a near-circular Earth orbit at an altitude of 100.4 nautical miles by 98.9 nautical miles.

After completing one and a half orbits around the Earth, the S-IVB stage reignited for the trans-lunar injection (TLI), which set Apollo 11 on its trajectory to the Moon. At 16:22:13 UTC, the TLI burn successfully propelled the spacecraft toward its destination. Roughly 30 minutes later, astronaut Michael Collins took control for the transposition, docking, and extraction maneuver, in which Columbia separated from the spent S-IVB stage, rotated 180 degrees, and docked with Eagle. Once the Lunar Module was successfully extracted, the combined spacecraft continued toward the Moon, while the S-IVB stage was slung into an orbit around the Sun.

On July 19, Apollo 11 passed behind the Moon and performed a service propulsion system (SPS) burn, slowing the spacecraft enough to enter lunar orbit at 17:21:50 UTC. Over the course of the next thirty orbits, the crew observed their landing site in the southern Sea of Tranquility, selected for its relatively flat and smooth terrain. This site had been thoroughly studied by previous lunar missions, including Ranger 8 and Surveyor 5, which provided detailed information about the area's surface conditions. The landing site was about 25 kilometers southeast of Surveyor 5's location and 68 kilometers southwest of Ranger 8's crash site, making it an ideal candidate for Apollo 11's historic landing.

As Eagle descended toward the lunar surface, Armstrong and Aldrin realized they were flying too fast, passing over landmarks seconds earlier than expected. This meant they would likely land several miles west of their intended target. The cause was possibly mascons—concentrations of high mass beneath the lunar surface that affect gravitational forces—or residual air pressure from the docking tunnel, but mission control could not be certain.

The Apollo 11 Lunar Module Eagle, in a landing
configuration was photographed in lunar orbit from the
Command and Service Module Columbia. Inside the
module were Commander Neil A. Armstrong and Lunar
Module Pilot Buzz Aldrin. The long rod-like protrusions
under the landing pods are lunar surface sensing probes.
Upon contact with the lunar surface, the probes sent a
signal to the crew to shut down the descent engine.

Five minutes into the descent, while Eagle was about 6,000 feet (1,800 meters) above the surface, the Lunar Module Guidance Computer (LGC) issued several 1201 and 1202 program alarms. These alarms signaled "executive overflows," meaning the computer was overloaded and postponing lower-priority tasks. In Mission Control, computer engineer Jack Garman assured Guidance Officer Steve Bales that the alarms were not critical and could be ignored, a message relayed to Armstrong and Aldrin. The computer was performing as designed, prioritizing essential functions and avoiding an abort, as later emphasized by Margaret Hamilton, Director of Apollo Flight Computer Programming at MIT. The problem stemmed from the rendezvous radar switch being in the wrong position, causing the computer to process data from both the landing and rendezvous radars simultaneously.

As Eagle neared its landing site, Armstrong looked outside and saw that the computer was guiding them toward a boulder-strewn area near a large crater, later identified as West Crater. Armstrong took semi-automatic control of the spacecraft, steering it manually to avoid the rough terrain. Aldrin continued calling out navigation data as Armstrong skillfully maneuvered Eagle over the hazardous area.

With only 90 seconds of fuel remaining, Armstrong guided the Lunar Module toward a clear patch of ground. Lunar dust kicked up by the descent engine made it difficult to see the surface, but Armstrong used the large rocks protruding from the dust cloud to gauge their speed. At 100 feet (30 meters) above the surface, Armstrong realized that their new landing site contained a small crater. He adjusted the descent, clearing the crater and steering to a level area beyond.

At just 67 inches (170 cm) above the surface, a light inside the Lunar Module indicated that one of Eagle's footpads had touched the ground. Aldrin radioed, "Contact light!"—the signal for Armstrong to shut off the engine, which he did three seconds later. Aldrin confirmed the spacecraft's safe landing by announcing, "Okay, engine stop. ACA—out of detent," referring to the Attitude Control Assembly. Armstrong acknowledged, and they completed the landing procedures.

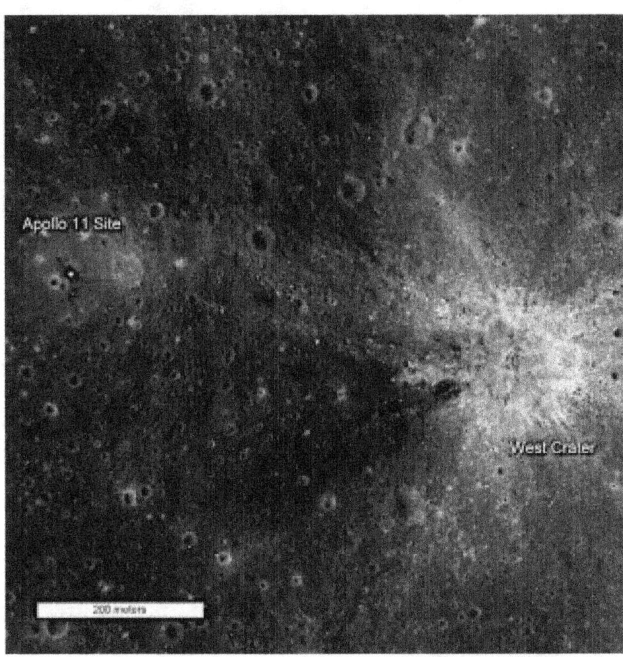

West Crater with boulders which Neil Armstrong had to
overfly before the first manned lunar landing. LROC
took this picture from orbit in autumn of 2009. The
descent stage of the Eagle is to the upper left.

At 20:17:40 UTC on Sunday, July 20, Eagle safely landed on the Moon with only 216 pounds (98 kg) of usable fuel remaining—enough for just 25 seconds of powered flight before an abort would have been necessary. Armstrong immediately radioed Mission Control, delivering the iconic message, "Houston, Tranquility Base here. The Eagle has landed." The change in call sign from "Eagle" to "Tranquility Base" signified the successful landing. CAPCOM Charlie Duke responded with relief: "Roger, Tranquility, we copy you on the ground. You got a bunch of guys about to turn blue. We're breathing again. Thanks a lot."

Two and a half hours after the historic landing, Aldrin broadcast a message to Earth, urging listeners to reflect on the significance of the moment. "I'd like to take this opportunity to ask every person listening in, whoever and wherever they may be, to pause for a moment and contemplate the events of the past few hours and to give thanks in his or her own way." Quietly, he took communion, using a kit prepared by his pastor at Webster Presbyterian Church in Houston. Aldrin, an elder at the church, performed the religious ceremony privately due to a lawsuit NASA faced from atheist activist Madalyn Murray O'Hair, who had objected to religious activities during Apollo 8's broadcast.

Although the astronauts were scheduled to rest after landing, they were too excited to sleep and decided to begin preparations for their historic EVA (extravehicular activity) earlier than planned. This decision would soon lead Armstrong and Aldrin to take humanity's first steps on another celestial body.

Preparations for the historic first lunar surface walk by Neil Armstrong and Buzz Aldrin began at 23:43 UTC, although the process took longer than expected, extending to three and a half hours rather than the planned two. On Earth, everything they needed had been meticulously organized in advance, but on the Moon, the cabin of the Lunar Module (LM) Eagle was cluttered with additional items such as checklists, food packets, and tools, complicating their tasks. Six hours and thirty-nine minutes after landing, Armstrong and Aldrin were finally ready to step outside, and the Eagle was depressurized.

At 02:39:33 UTC, the hatch of the Eagle was opened. Armstrong, wearing his Portable Life Support System (PLSS), initially found it difficult to squeeze through the hatch, highlighting the physical challenges astronauts faced during the Lunar Module's egress and ingress. With heart rates elevated due to the effort, Armstrong descended the nine-rung ladder, his chest-mounted control unit blocking his view of his feet. As he descended, he deployed the Modular Equipment Stowage Assembly (MESA), activating a TV camera to capture the monumental event.

The Apollo 11 mission used a slow-scan television system incompatible with broadcast TV, necessitating a workaround where the images were displayed on a monitor and re-recorded for public broadcast, reducing picture quality. Despite this limitation, the live transmission was received by Honeysuckle Creek Tracking Station in Australia, followed by the Parkes Radio Telescope, which captured the broadcast transmitted to over 600 million viewers worldwide.

At 02:56:15 UTC, six and a half hours after landing, Armstrong descended from the Eagle and set foot on the Moon. His words—"That's one small step for [a] man, one giant leap for mankind"—became one of the most iconic statements in history, despite the controversy over whether the article "a" was spoken or lost in transmission. A digital analysis of the recording later suggested that the "a" might have been said but obscured by static, though Armstrong himself later admitted he may have misspoken.

Soon after, Armstrong collected a contingency soil sample as a safeguard in case the mission had to be cut short, ensuring that at least some lunar material would be brought back to Earth. Following this, he removed the TV camera from the MESA, mounted it on a tripod, and made a panoramic sweep of the lunar surface. Aldrin joined him shortly afterward, stepping onto the Moon with the words, "Magnificent desolation."

The astronauts quickly adapted to the Moon's gravity, which was one-sixth that of Earth. Armstrong described movement as surprisingly easy, though the astronauts found themselves having to plan several steps ahead due to the fine, slippery lunar soil. Aldrin tested various methods of movement, including kangaroo-like hops, though both astronauts noted that their PLSS backpacks caused a tendency to tip backward. The duo spent time adjusting to these challenges while performing their designated tasks.

One of their most symbolic acts on the Moon was the planting of the American flag. Despite some difficulty with the telescoping rod, which penetrated only two inches into the lunar surface, the flag was raised and left standing. Aldrin, with his military precision, gave a crisp salute while Armstrong took photographs. At that moment, President Richard Nixon made a direct call to the astronauts from the White House, calling their achievement "the most historic telephone call ever made." Armstrong humbly responded, emphasizing that they represented not just the United States but all of humanity.

Aldrin salutes the deployed United States flag on the lunar surface.

As the astronauts worked on the Moon, they deployed the Early Apollo Scientific Experiments Package (EASEP), which included a Passive Seismic Experiment Package (PSEP) to measure moonquakes and a retroreflector array for the lunar laser ranging experiment. Armstrong ventured 196 feet from the LM to photograph the rim of Little West Crater, while Aldrin used a geologist's hammer to collect core samples. Although Aldrin struggled to penetrate the lunar surface deeply, they gathered rock and soil samples, uncovering new minerals, including armalcolite, tranquillityite, and pyroxferroite—significant geological discoveries.

In a lasting testament to their journey, Armstrong unveiled a plaque on the Eagle's ladder that read, "Here men from the planet Earth first set foot upon the Moon, July 1969, A.D. We came in peace for all mankind." The plaque, signed by the astronauts and President Nixon, served as a message to future explorers, commemorating the peaceful intentions of this pioneering mission. As their time on the lunar surface neared its end, Armstrong's metabolic rates prompted a coded message from Mission Control urging him to slow down. Despite these warnings, both astronauts maintained generally lower-than-expected metabolic rates, allowing Mission Control to grant them a 15-minute extension for additional tasks. In later interviews, Armstrong explained that NASA had limited the time and distance of this first moonwalk due to uncertainties about how much cooling water the PLSS backpacks would consume as the astronauts generated body heat while working.

Apollo 11 Lunar Module Pilot Buzz Aldrin's bootprint.
Aldrin photographed this bootprint about an hour into
their lunar extra-vehicular activity on July 20, 1969, as
part of investigations into the soil mechanics of the lunar
surface. This photo would later become synonymous
with humankind's venture into space.

With the successful completion of the first lunar EVA, Armstrong and Aldrin had not only collected valuable scientific data but also left a profound legacy that continues to inspire humanity's quest for exploration beyond Earth.

After completing their historic exploration of the lunar surface, Buzz Aldrin was the first to re-enter the Eagle. With some difficulty, the astronauts used a flat cable pulley device known as the Lunar Equipment Conveyor (LEC) to lift film and two sample boxes containing 21.55 kilograms (47.5 pounds) of lunar material up to the Lunar Module (LM) hatch. The LEC proved cumbersome, leading future Apollo missions to abandon its use in favor of carrying equipment and samples by hand. Once inside, Armstrong reminded Aldrin of a bag of memorial items stored in his sleeve pocket, which Aldrin tossed down to the surface before Armstrong climbed up the ladder and re-entered the LM.

Inside the LM, the astronauts transferred to its life support system and lightened the ascent stage in preparation for their return to lunar orbit. They discarded their Portable Life Support System (PLSS) backpacks, lunar overshoes, an empty Hasselblad camera, and other equipment. At 05:11:13 UTC, they closed the hatch, repressurized the cabin, and settled down to sleep after a momentous day.

Astronaut Buzz Aldrin on the moon Full description: Astronaut Buzz Aldrin, lunar module pilot, stands on the surface of the moon near the leg of the lunar module, Eagle, during the Apollo 11 moonwalk. Astronaut Neil Armstrong, mission commander, took this photograph with a 70mm lunar surface camera

In the event that something went wrong and the astronauts were stranded on the Moon, presidential speechwriter William Safire had prepared a solemn announcement for President Richard Nixon to deliver. The speech, titled "In Event of Moon Disaster," detailed a protocol to be followed in such a tragedy. Mission Control would "close down communications" with the astronauts, and a clergyman would commend their souls in a public ceremony akin to a burial at sea. The prepared speech included an allusion to Rupert Brooke's World War I poem "The Soldier," a poignant reminder of the risks involved in such an ambitious mission. Although Michael Collins remained in lunar orbit aboard the Command Module Columbia, the speech did not reference him, as it was expected that Collins would return to Earth in the event of a mission failure.

Inside the cabin, Aldrin accidentally broke the circuit breaker that armed the main engine for liftoff, a critical component for their return to lunar orbit. The astronauts feared this would prevent them from leaving the Moon. Ingeniously, Aldrin used the non-conductive tip of a felt-tip pen to activate the switch, averting what could have been a catastrophic situation.

Before departing the lunar surface, the astronauts left behind several memorial items: an Apollo 1 mission patch in honor of Roger Chaffee, Gus Grissom, and Ed White, the astronauts who perished in a cabin fire during a pre-launch test in 1967; two commemorative medals honoring Soviet cosmonauts Vladimir Komarov and Yuri Gagarin, both of whom died in service to space exploration; a gold olive branch, symbolizing peace; and a silicon disk inscribed with goodwill messages from 73 world leaders, as well as statements from U.S. presidents Eisenhower,

Kennedy, Johnson, and Nixon. The disk also listed the leadership of the U.S. Congress, members of the committees responsible for NASA legislation, and NASA's top management.

After resting for about seven hours, the astronauts were awakened by Mission Control to prepare for their return flight. At that moment, unknown to them, the Soviet probe Luna 15—which had been orbiting the Moon in parallel with Apollo 11—was attempting its descent in a last-minute effort to return lunar samples before the Americans. However, just two hours before the planned liftoff of Eagle, Luna 15 crashed on the lunar surface, bringing an end to a dramatic chapter in the Space Race. British astronomers monitoring the mission remarked on the high stakes of this cosmic drama, acknowledging the significance of this culmination in human history.

At 17:54:00 UTC on July 21, 1969, the ascent stage of Eagle lifted off from the lunar surface, carrying Armstrong and Aldrin back to join Collins aboard Columbia in lunar orbit. As the ascent stage fired, the American flag planted near the LM was caught in the exhaust, whipping violently before toppling over. Aldrin, focusing on the spacecraft's computers, glanced up just in time to see the flag fall. Future Apollo missions planted their flags farther from the Lunar Module to avoid this mishap.

With their successful liftoff and rendezvous in lunar orbit, the astronauts were now prepared for the final leg of their journey—returning home to Earth.

During his solo flight around the Moon aboard the Command Module Columbia, Michael Collins never felt isolated or lonely, despite popular descriptions of his experience as "not since Adam has any human known such solitude." In his autobiography, Collins emphasized his integral role in the mission, writing, "This venture has been structured for three men, and I consider my third to be as necessary as either of the other two." Although he spent 48 minutes of each orbit out of radio contact with Earth while passing around the far side of the Moon, Collins described his feelings not as fear or loneliness, but as "awareness, anticipation, satisfaction, confidence, almost exultation."

One of Collins' primary tasks during his time in lunar orbit was to locate the Eagle Lunar Module on the surface. Mission Control radioed to inform him that they believed the Eagle had landed approximately 4 miles (6.4 kilometers) off the intended target. Each time Collins passed over the estimated landing site, he searched in vain for the module. Despite repeated efforts, he was unable to visually locate the Eagle on the lunar surface during his early orbits.

In addition to his search for the Eagle, Collins performed essential maintenance tasks aboard Columbia. These included dumping excess water produced by the fuel cells and preparing the cabin for Armstrong and Aldrin's eventual return. On his third orbit around the Moon, just before entering the far side's radio blackout, Mission Control informed him of a potential issue with the temperature of the coolant system. If the temperature dropped too low, there was a risk that parts of the spacecraft might freeze. Collins, demonstrating calm and methodical decision-making, chose to switch the system from automatic to manual and back to automatic again, a simple adjustment that resolved the issue without requiring the more complex Environmental Control System Malfunction Procedure 17. As Columbia returned to the near side of the Moon, he reported to Mission Control that the issue had been successfully resolved.

The Apollo 11 Command and Service Modules (CSM) are photographed from the Lunar Module (LM) in lunar orbit during the Apollo 11 lunar landing mission. The lunar surface below is in the north central Sea of Fertility. The coordinates of the center of the picture are 51 degrees east longitude and 1 degree north latitude. About half of the crater Taruntius G is visible in the lower left corner of the picture. Part of Taruntius H can be seen at lower right.

For the next few orbits, Collins described his time on the far side of the Moon as "relaxing," a brief period of quiet solitude amidst the intense activities of the mission. After Armstrong and Aldrin completed their historic moonwalk, Collins took time to rest in preparation for the critical rendezvous with the Eagle. While the mission plan called for the Eagle to return and dock with Columbia in lunar orbit, Collins was fully prepared for a contingency where he would need to fly Columbia down to meet the Eagle if necessary.

Despite the challenges and technical complexities, Collins' time in lunar orbit was marked by his steady composure and unwavering confidence. His role, while often less visible than that of Armstrong and Aldrin, was indispensable to the mission's success. His careful management of Columbia ensured a smooth and safe reunion with the Eagle, completing the lunar phase of the Apollo 11 mission and paving the way for the crew's return to Earth.

The Eagle's ascent stage successfully rendezvoused with the Command Module Columbia at 21:24 UTC on July 21, 1969. Docking occurred 11 minutes later, at 21:35 UTC, reuniting Neil Armstrong and Buzz Aldrin with Michael Collins, who had been orbiting the Moon in Columbia. With their mission on the lunar surface complete, the astronauts jettisoned the Eagle's ascent stage into lunar orbit at 23:41 UTC. Initially, it was thought that the ascent stage remained in lunar orbit, but later reports indicated its orbit decayed, causing it to crash into the Moon at an unknown location. However, some 2021 calculations suggested that the Eagle ascent stage might still be orbiting the Moon.

As the astronauts neared their return to Earth, they made a final television broadcast on July 23, the night before their splashdown. Michael Collins reflected on the flawless performance of the Saturn V rocket, the vehicle that had carried them into space, praising the thousands of people whose hard work made the mission possible. In his heartfelt message, Collins remarked:

"... The Saturn V rocket which put us in orbit is an incredibly complicated piece of machinery, every piece of which worked flawlessly ... We have always had confidence that this equipment will work properly. All this is possible only through the blood, sweat, and tears of a number of people ... All you see is the three of us, but beneath the surface are thousands and thousands of others, and to all of those, I would like to say, 'Thank you very much.'"

Buzz Aldrin followed with a broader reflection, noting that the mission transcended individual and national efforts, symbolizing humanity's drive to explore the unknown. He quoted from the Bible's Book of Psalms, saying:

"This has been far more than three men on a mission to the Moon; more, still, than the efforts of a government and industry team; more, even, than the efforts of one nation. We feel that this stands as a symbol of the insatiable curiosity of all mankind to explore the unknown ... Personally, in reflecting on the events of the past several days, a verse from Psalms comes to mind. 'When I consider the heavens, the work of Thy fingers, the Moon and the stars, which Thou hast ordained; What is man that Thou art mindful of him?'"

Neil Armstrong closed the broadcast with a message of gratitude to the American people and the thousands who contributed to the mission's success. He recognized the giants of science and the determination of four U.S. administrations that had made the Apollo program possible:

"The responsibility for this flight lies first with history and with the giants of science who have preceded this effort; next with the American people, who have, through their will, indicated their desire; next with four administrations and their Congresses, for implementing that will; and then, with the agency and industry teams that built our spacecraft, the Saturn, the Columbia, the Eagle, and the little EMU, the spacesuit and backpack that was our small spacecraft out on the lunar surface. We would like to give special thanks to all those Americans who built the spacecraft; who did the construction, design, the tests, and put their hearts and all their abilities into those craft. To those people tonight, we give a special thank you, and to all the other people that are listening and watching tonight, God bless you. Good night from Apollo 11."

During their return to Earth, the mission faced one last technical hurdle when a bearing at the Guam tracking station failed, potentially compromising communication during the final re-entry phase. With little time for a regular repair, station director Charles Force enlisted the help of his ten-year-old son, Greg, whose small hands were able to reach into the housing and pack the bearing with grease. Thanks to this quick thinking, communication was maintained, and Armstrong later personally thanked Greg for his assistance.

Columbia floats on the ocean as Navy divers assist in
retrieving the astronauts.

On July 24, 1969, Apollo 11's Columbia command module splashed down in the Pacific Ocean, concluding humanity's first successful mission to land on the Moon. The spacecraft, carrying astronauts Neil Armstrong, Buzz Aldrin, and Michael Collins, was recovered by the USS Hornet, the primary recovery ship. Preparations for the retrieval began well before the splashdown, as NASA had selected the Hornet, under the command of Captain Carl J. Seiberlich, on June 5, 1969, to take over recovery duties from the USS Princeton, which had supported Apollo 10. The Hornet was quickly modified for the mission, leaving its home port in Long Beach, California, and arriving in Pearl Harbor on July 5. There, it embarked specialized recovery teams, media representatives, and key recovery equipment, including a boilerplate command module for training.

On July 12, the Hornet left Pearl Harbor and sailed to the recovery area in the central Pacific, near 10°36′N 172°24′E, while Apollo 11 remained on the launch pad at Kennedy Space Center. A presidential party, including President Richard Nixon, NASA Administrator Thomas Paine, and several dignitaries, traveled separately to Johnston Atoll and eventually to the recovery site aboard the command ship USS Arlington. Their mission was to witness the astronauts' return and celebrate the historic achievement.

As the recovery teams prepared, concerns arose regarding the weather. US Air Force Captain Hank Brandli, using classified satellite imagery, discovered that a storm front threatened the splashdown zone. The possibility of strong winds and poor visibility posed a serious risk to the mission. Brandli relayed this information to Navy Captain Willard S. Houston Jr., who advised NASA to move the recovery location 215 nautical miles northeast, a decision that required altering the re-entry flight plan. This unprecedented change involved using a never-before-attempted sequence of computer programs for the re-entry process. The astronauts would experience intense deceleration forces, but the adjustment ensured their safe return.

On July 24, before dawn, four Sikorsky SH-3 Sea King helicopters and three Grumman E-1 Tracer aircraft launched from the Hornet to track and assist in the recovery operation. At 16:44 UTC, Columbia's drogue parachutes deployed, signaling the final moments of re-entry. Seven minutes later, the command module struck the water approximately 24 kilometers from the Hornet. Despite landing upside down, the astronauts activated flotation bags to right the spacecraft. Navy divers deployed from the helicopters and attached flotation collars and a sea anchor to stabilize the module.

Due to concerns about potential lunar pathogens, the astronauts donned biological isolation garments (BIGs) before being helped into a recovery raft. They were rubbed down with sodium hypochlorite solution to decontaminate any lunar dust, and Columbia was similarly cleaned with Povidone-iodine. Once the astronauts were winched aboard the recovery helicopter, they were flown back to the Hornet, where they entered the Mobile Quarantine Facility (MQF). This isolation protocol, part of NASA's quarantine measures for extraterrestrial exposure, required the astronauts to spend 21 days in quarantine, ensuring that no harmful organisms from the Moon were brought back to Earth.

After a brief ceremony with President Nixon, who greeted the astronauts from outside the MQF, the Hornet sailed back to Pearl Harbor. From there, Columbia and the MQF were flown to the Manned Spacecraft Center in Houston. The astronauts remained in quarantine until they received a clean bill of health on August 10, 1969, officially ending the period of concern over lunar contamination.

The world's response to Apollo 11's success was overwhelming. On August 13, 1969, New York City and Chicago hosted ticker-tape parades attended by millions, and that evening, a state dinner was held in Los Angeles, where Nixon awarded the astronauts the Presidential Medal of Freedom. The astronauts later addressed a joint session of Congress, presenting two American flags they had carried to the lunar surface. These events kicked off a 38-day world tour that took the Apollo 11 crew to 22 countries, including Mexico, Brazil, Spain, France, and Japan. They were celebrated globally as pioneers, and their accomplishment became a symbol of human achievement and exploration.

The Apollo 11 Mobile Quarantine Facility

Many nations honored the landing by issuing commemorative stamps and coins, forever marking the moment when humanity first set foot on another world. Apollo 11's success resonated worldwide, a triumph of technology, perseverance, and the human spirit, as the Space Race reached its pinnacle.

The Apollo 11 mission left an indelible mark on both American culture and the world at large, symbolizing the achievement of President John F. Kennedy's goal set nearly a decade earlier. On July 20, 1969, as Neil Armstrong took his first steps on the lunar surface, Mission Control in Houston displayed the phrase, "TASK ACCOMPLISHED, July 1969," signifying not just a victory for the Apollo program but for the United States in the broader context of the Cold War. The successful Moon landing underscored American technological superiority and effectively declared that the United States had won the Space Race against the Soviet Union.

The cultural impact of the Apollo 11 landing was immediate and widespread. New expressions entered the English language, most notably the phrase, "If they can send a man to the Moon, why can't they...?" Armstrong's iconic words, "That's one small step for [a] man, one giant leap for mankind," were immortalized in popular culture and became the subject of countless parodies and references in media. Yet, while many celebrated the achievement, not everyone shared in the jubilation. The space program's immense cost was seen by some as emblematic of the social and economic disparities within the United States, particularly by marginalized communities.

Ticker tape parade for the Apollo 11 astronauts. Location is Manhattan, New York City on the section of Broadway known as the "Canyon of Heroes". Pictured in the lead car, from the right, are astronauts Neil A. Armstrong, Michael Collins and Edwin E. Aldrin, Jr.

On the eve of the Apollo 11 launch, civil rights leader Ralph Abernathy led a group of protesters outside Kennedy Space Center. They expressed concerns about the nation's priorities, pointing to poverty and racial inequality as critical issues that demanded attention. NASA Administrator Thomas Paine met with Abernathy and acknowledged the protest, even inviting the group to view the historic launch. Abernathy, moved by the spectacle, prayed for the astronauts' safety, but his protest highlighted the contrast between technological achievement and the struggles many Americans faced at home. This sentiment was captured in Gil Scott-Heron's 1970 poem "Whitey on the Moon," which criticized the expenditure on space exploration while neglecting pressing social needs.

The landing captivated global audiences, with an estimated 20% of the world's population watching the historic event unfold. Yet, as Apollo 11 completed its mission, public interest in the Apollo program began to wane. Subsequent lunar missions did not hold the same appeal. While landing a person on the Moon was an easily comprehensible goal, lunar geology and other scientific pursuits were more abstract and failed to resonate with the broader public. With Kennedy's objective achieved, many questioned the value of continuing the lunar missions, and support for space exploration diminished.

Public opinion reflected this shift. Throughout the late 1960s, only once did a majority of Americans, according to the Gallup Poll, support increased spending on space exploration. By 1973, as inflation rose and Cold War tensions eased, 59% of Americans favored cutting the space budget. Although the Apollo program had delivered a monumental achievement, the economic pressures of the time prompted government officials to scale back. Caspar Weinberger, then deputy director of the Office of Management and Budget, warned that drastic cuts to the space program might suggest that the nation's greatest accomplishments were behind it.

The response from the Soviet Union to Apollo 11 was notably muted. Soviet officials publicly downplayed the significance of the Moon landing, arguing that it was unnecessary and too dangerous. At the time, the Soviets were focused on retrieving lunar samples using robotic missions. While the government officially denied that they had been racing to the Moon, it was later revealed that the Soviet Union had indeed attempted to develop a manned lunar mission but had been thwarted by technological challenges. The Soviet public's reaction to the Apollo 11 landing was mixed, with limited coverage in the media contributing to a subdued response.

Apollo 11's legacy extended into popular culture, inspiring songs such as "Armstrong, Aldrin and Collins" by the Byrds, "Coon on the Moon" by Howlin' Wolf, and "One Small Step" by Ayreon. These musical tributes, along with numerous references in film, literature, and art, ensured that the story of the first human landing on the Moon would remain a cultural touchstone for generations to come.

Though the excitement of the Apollo era began to fade in the years following the Moon landing, Apollo 11's success demonstrated what humanity could achieve when driven by a clear, ambitious goal. It was more than just a technical triumph; it was a defining moment in human history, a reminder that exploration and discovery are intrinsic to the human spirit.

The Apollo 11 Command Module "Columbia" today after artifact cleaning

Chapter 6 - Apollo's Legacy and Beyond

Apollo 12

First precisely targeted piloted landing on the Moon (Surveyor 3 site)

Apollo 12, launched on November 14, 1969, from Kennedy Space Center, marked the sixth crewed flight in NASA's Apollo program and the second mission to successfully land on the Moon. The crew was led by Commander Charles "Pete" Conrad, with Alan L. Bean as the Lunar Module Pilot and Richard F. Gordon serving as the Command Module Pilot. This mission, lasting from November 14 to November 24, 1969, was designed not only to continue the exploration of the lunar surface but also to demonstrate the ability to perform a pinpoint landing on a specific target on the Moon.

Originally, Apollo 12 was intended to be the backup mission for Apollo 11, should Neil Armstrong's historic landing attempt have failed. However, after the success of Apollo 11, NASA postponed Apollo 12 by two months, allowing for a more relaxed schedule and additional training. In preparation, Conrad and Bean were given extended geological field training, enabling them to conduct more thorough scientific exploration of the lunar surface. Their spacecraft and launch vehicle were nearly identical to those of Apollo 11, with one notable improvement: hammocks installed inside the Lunar Module to allow the astronauts to rest more comfortably during their stay on the Moon.

The landing site for Apollo 12 was a critical choice for NASA. While the site selection for Apollo 11 was primarily dictated by safety and the need for a relatively flat and hazard-free area near the lunar equator, scientific interest played a larger role in the decision for Apollo 12. Initially, NASA planned for the mission to land in Sinus Medii, just west of the Sea of Tranquility, where Apollo 11 had touched down. However, mission planners soon proposed a landing near the Surveyor 3 probe, which had arrived on the Moon in 1967. This location offered the opportunity to conduct lunar exploration and retrieve parts from the unmanned Surveyor probe, providing insight into how materials fared after exposure to the harsh lunar environment.

While some within NASA were concerned about the risk of missing the Surveyor landing site due to the challenges of precision landing, the ability to land accurately was deemed essential for future missions. On July 25, 1969, Apollo Program Manager Samuel Phillips approved the landing site near Surveyor crater, despite opposition from some members of the site selection boards. This decision set the stage for the mission's successful landing, which became a crucial demonstration of the Apollo program's evolving capabilities.

Commander, Charles "Pete" Conrad Jr.; Command
Module pilot, Richard F. Gordon Jr.; and Lunar Module
pilot, Alan L. Bean

The Apollo 12 crew was an all-Navy team, featuring Commander Charles "Pete" Conrad, Command Module Pilot Richard "Dick" Gordon, and Lunar Module Pilot Alan L. Bean. Conrad, 39 years old at the time of the mission, had a distinguished career as a naval aviator and test pilot before joining NASA's second astronaut group in 1962. He had previously flown on Gemini 5 and commanded Gemini 11. Gordon, 40 years old, also had a strong naval background, having graduated with a degree in chemistry from the University of Washington and completed test pilot school. He had flown alongside Conrad on Gemini 11. Bean, 37 years old, had been a student of Conrad's at the United States Naval Test Pilot School and was selected as an astronaut in 1963. He replaced Clifton C. Williams, who tragically died in a plane crash in 1967, after Conrad requested him for the mission.

In addition to the prime crew, the Apollo 12 mission had a backup crew consisting of David R. Scott as Commander, Alfred M. Worden as Command Module Pilot, and James B. Irwin as Lunar Module Pilot. This team would later be assigned to Apollo 15. The support crew, a group created to assist in managing the mission's vast logistical needs, included astronauts Gerald P. Carr, Edward G. Gibson, and Paul J. Weitz. These support astronauts were responsible for assembling mission rules, flight plans, and checklists, ensuring everything was in order for the prime crew's mission.

Mission control was also critical to the success of Apollo 12. Four flight directors, including Gerry Griffin, Pete Frank, Clifford E. Charlesworth, and Milton Windler, oversaw the mission from different shifts. Their job, succinctly described as "taking any actions necessary for crew safety and mission success," placed them at the helm of one of NASA's most complex endeavors. Capsule communicators (CAPCOMs), including the backup crew members and others like Don Lind, provided real-time communication with the crew in space, ensuring that every aspect of the mission was carefully coordinated.

Apollo 12's lunar surface activities focused on scientific exploration and sample collection, as well as the retrieval of components from the Surveyor 3 probe. The astronauts spent over 31 hours on the lunar surface, deploying scientific instruments and conducting experiments as part of the Apollo Lunar Surface Experiments Package (ALSEP). Their landing near Surveyor 3 allowed them to retrieve parts of the probe, which were returned to Earth for analysis, revealing how materials fared after two years on the Moon. The mission's success in landing within walking distance of Surveyor 3 proved the accuracy of the Apollo program's landing capabilities and set the stage for more ambitious lunar explorations in subsequent missions.

The training and preparation for Apollo 12 were meticulous and exhaustive, reflecting NASA's growing experience and ambition in the Apollo program. Each Apollo 12 astronaut underwent over 1,000 hours of mission-specific training, exceeding the preparation for Apollo 11. This was in addition to the 1,500 hours of general training they had already completed as backup crew members for Apollo 9. The intensive schedule included more than 400 hours in simulators for the Command Module (CM) and Lunar Module (LM), some of which were linked in real-time to Mission Control to simulate real mission scenarios. Conrad, the mission commander, also flew the Lunar Landing Training Vehicle (LLTV), despite a previous near-disaster where Neil Armstrong had to eject from a similar vehicle before it crashed in 1968.

One of the defining aspects of the preparation was the focus on lunar surface activities. After the high-profile nature of Apollo 11, Commander Conrad requested that the geology training be conducted without media involvement, preferring a more focused approach. The astronauts' field trips to practice sample collection were kept low-key, with communications maintained with a Capsule Communicator (CAPCOM) and geologists in a nearby tent. These field exercises simulated the lunar environment as closely as possible, with the astronauts collecting and photographing samples just as they would on the Moon. However, frustrations arose when the documentation procedures were repeatedly changed by scientists, leading Conrad to insist on a final set of instructions. Additionally, the Apollo 12 crew benefited from the opportunity to study lunar samples returned by Apollo 11, further refining their geological skills.

For Apollo 12, site selection was more precise than for Apollo 11. While Apollo 11 had targeted a large, ellipse-shaped landing zone, Apollo 12 was planned to land near the Surveyor 3 probe, which had arrived on the Moon in 1967. This required a pinpoint landing, a skill NASA needed to develop for future missions. Geology traverses were planned before the mission, marking the first time astronauts would have predefined routes for lunar exploration, a practice that would become essential in later missions.

The stages of Apollo 12's Lunar Module, LM-6, were delivered to Kennedy Space Center (KSC) on March 24, 1969, followed by the Command and Service Modules (CSM-108) on March 28. The spacecraft was mated with its launch vehicle, Saturn V SA-507, and rolled out to Launch Complex 39A on September 8, 1969. As with Apollo 11, the Saturn V launch vehicle had no significant changes, though 17 additional instrumentation measurements were added, bringing the total to 1,365. The entire vehicle, including the spacecraft, weighed 6,487,742 pounds at launch, slightly heavier than Apollo 11. The Command Module and Lunar Module had only minor modifications from their predecessors, such as the addition of hammocks to allow the astronauts to rest more comfortably on the Moon.

One of the most significant aspects of the Apollo 12 mission was the third stage trajectory. After the Lunar Module separated from the Saturn V, the third stage, known as the S-IVB, was intended to fly into solar orbit. However, due to a miscalculation, the S-IVB missed achieving Earth escape velocity and instead remained in a semi-stable Earth orbit until 1971. Decades later, it briefly returned to Earth orbit in 2002 before being identified as an artificial object by amateur astronomer Bill Yeung. The S-IVB has since returned to solar orbit, with projections indicating it may be recaptured by Earth's gravity sometime in the 2040s. In later Apollo missions, the S-IVBs were deliberately crashed into the Moon to generate seismic data for scientific study.

The Apollo 12 spacecraft was composed of Command and Service Modules (CSM-108) and Lunar Module 6 (LM-6), along with a Launch Escape System (LES) and Spacecraft-Lunar Module Adapter (SLA-15). The LES was equipped with three rocket motors designed to propel the Command Module to safety in the event of an emergency during launch. Meanwhile, the SLA housed the Lunar Module and provided a structural connection between the Saturn V and the LM. The Launch Escape System was modified slightly from Apollo 11 with a more reliable motor igniter, but the SLA remained unchanged.

The Command Module for Apollo 12 was given the call sign Yankee Clipper, and the Lunar Module was dubbed Intrepid. The all-Navy crew selected these sea-related names from suggestions provided by employees of the spacecraft's prime contractors. Yankee Clipper represented the pride and prestige of American ships, while Intrepid symbolized the astronauts' fortitude and determination in exploring space.

There were a few modifications to the Command Module's systems, such as the addition of a hydrogen separator to prevent gaseous hydrogen from entering the potable water tank—a problem that had caused discomfort for the Apollo 11 crew. Additionally, the recovery system was strengthened so that an auxiliary loop would no longer need to be attached by recovery swimmers after splashdown. The Lunar Module was also modified slightly to allow it to carry scientific experiment packages for deployment on the lunar surface. Notably, Apollo 12 featured a color television camera, replacing the black-and-white camera used on Apollo 11, offering clearer images of the lunar surface and the astronauts' activities.

Apollo 12's preparation, hardware, and spacecraft demonstrated NASA's growing confidence and capability in lunar exploration. From the enhanced training to the fine-tuned spacecraft systems, the mission was a carefully orchestrated step forward in expanding human presence on the Moon and collecting valuable scientific data.

On November 14, 1969, Apollo 12 launched from Kennedy Space Center at 11:22 a.m. Eastern Standard Time (16:22:00 UT), marking a historic moment as President Richard Nixon became the first sitting U.S. president to witness a crewed space launch. Vice President Spiro Agnew was also present, underscoring the significance of the mission for both NASA and the U.S. government. The launch, planned to reach the Moon under optimal lighting conditions, took place under completely overcast skies, with the spacecraft encountering winds of up to 151.7 knots (174.6 mph), the strongest recorded for any Apollo mission during ascent.

Despite the challenging weather, including the risk of cumulonimbus clouds, NASA decided to proceed with the launch, waiving a rule against launching through such conditions. The window for Apollo 12's mission allowed for a delay of up to two days, but with Apollo 11's success removing the pressure to achieve the first lunar landing, NASA could have waited until December for the next opportunity. However, with all systems go, the countdown proceeded, and Apollo 12 roared into the sky.

Thirty-six and a half seconds after liftoff, lightning struck the Saturn V rocket. The massive static discharge, triggered by the vehicle itself as it ascended through the stormy skies, knocked out all three of the spacecraft's fuel cells. A second lightning strike at 52 seconds caused further complications, disabling the "8-ball" attitude indicator, which displayed the spacecraft's orientation. Telemetry from the spacecraft to Mission Control was garbled, leaving the astronauts with a red board full of caution and warning lights. Unbeknownst to the crew, however, the Saturn V rocket itself was unaffected; its guidance system, controlled by the instrument unit, continued to function normally, allowing the rocket to remain on course.

In the midst of the chaos, John Aaron, a flight controller specializing in the spacecraft's electrical and environmental systems, recognized the telemetry pattern from a previous test where power loss had affected the signal conditioning electronics (SCE) that converted raw instrumentation data for display in Mission Control. With quick thinking, Aaron made the call, "Flight, EECOM. Try SCE to Aux," directing the crew to switch the SCE to a backup power supply. While the switch was relatively obscure, Lunar Module Pilot Alan Bean knew where to find it and successfully flipped the switch, restoring telemetry to Mission Control. Once the telemetry returned, it revealed that no major systems had been critically damaged. Bean quickly brought the fuel cells back online, and the mission was able to continue as planned.

Once the spacecraft reached Earth parking orbit, the crew conducted a thorough check of their systems before reigniting the Saturn V's S-IVB third stage for trans-lunar injection, sending them on their way to the Moon. Despite the initial electrical failure caused by the lightning strikes, the spacecraft showed no lasting damage, and the mission pressed forward.

One lingering concern after the lightning strike was the potential damage to the explosive bolts in the Command Module's parachute compartment, critical for opening the compartment at the end of the mission. The fear was that the lightning might have rendered the parachutes unusable, a malfunction that could only be discovered upon reentry. However, the decision was made not to inform the astronauts of this risk. Whether returning from Earth orbit or from the Moon, failed parachutes would mean certain death, and aborting the mission would not change the outcome. The crew, unaware of this concern, proceeded with their mission to the Moon.

Following the harrowing lightning strikes during launch, the Apollo 12 crew conducted careful systems checks while in Earth orbit. Once everything was confirmed to be functioning properly, the spacecraft executed the trans-lunar injection burn using the S-IVB stage. This critical maneuver occurred at exactly 02:47:22.80 into the mission, propelling Apollo 12 toward the Moon. Approximately one hour and twenty minutes later, the Command and Service Module (CSM) separated from the S-IVB stage. Command Module Pilot Richard Gordon then performed the transposition, docking, and extraction maneuver, successfully docking the CSM with the Lunar Module (LM) and extracting the combined craft from the S-IVB. The stage was then sent on a trajectory to attempt entry into solar orbit—a departure from Apollo 11, where the Service Module's propulsion system had been used to move away from the S-IVB stage.

Due to the lightning strikes experienced during launch, there were concerns that the Lunar Module might have been damaged. As a precaution, Commander Charles "Pete" Conrad and Lunar Module Pilot Alan Bean entered the LM earlier than scheduled on the first day of the flight to inspect its systems. Fortunately, no damage was found, and the mission proceeded as planned. The spacecraft made a single midcourse correction at 30 hours, 52 minutes, and 44.36 seconds into the mission, which placed Apollo 12 on a hybrid, non-free-return trajectory. This was a significant departure from previous crewed lunar missions, which had relied on a free-return trajectory—allowing the spacecraft to return to Earth without further propulsion if engine failure occurred. Instead, the hybrid trajectory required an additional burn to return to Earth, which could be performed using the LM's Descent Propulsion System (DPS) if necessary. This trajectory provided more flexibility in mission planning, allowing for a daylight launch and precise landing on the Moon. However, it extended the journey from trans-lunar injection to lunar orbit by an additional eight hours compared to previous missions.

Apollo 12 entered lunar orbit at 83 hours, 25 minutes, and 26.36 seconds into the mission. The Service Propulsion System (SPS) executed a burn lasting 352.25 seconds, placing the spacecraft into an elliptical orbit with an altitude of 170.2 by 61.66 nautical miles. On the first pass around the Moon, the crew transmitted a live television broadcast, providing viewers back on Earth with high-quality video of the lunar surface. By the third orbit, the SPS fired again to circularize the spacecraft's orbit at 66.1 by 54.59 nautical miles, preparing for the lunar landing.

At 107 hours, 54 minutes, and 2.3 seconds, the Command and Service Module (CSM) undocked from the Lunar Module (LM). Thirty minutes later, the CSM executed a burn to separate from the LM, placing the two vehicles 2.2 nautical miles apart. The Lunar Module, named Intrepid, began its descent toward the Moon with a 29-second burn at 109 hours, 23 minutes, and 39.9 seconds, moving into a lower orbit. The powered descent to the lunar surface commenced at 110 hours, 20 minutes, and 38.1 seconds.

As Intrepid descended, Conrad eagerly searched for a distinctive pattern of craters known as "The Snowman," which would confirm they were on the correct landing path. To his relief, the Snowman formation appeared exactly where it should be, signaling that the spacecraft was precisely on course. Conrad manually controlled the final descent, aiming for an area near the Surveyor crater dubbed "Pete's Parking Lot." However, the terrain proved rougher than anticipated, requiring additional maneuvering. Finally, at 110 hours, 32 minutes, and 36.2 seconds (06:54:36

UT on November 19, 1969), Intrepid touched down just 535 feet from the Surveyor 3 probe—successfully

The Apollo 12 lunar module Intrepid prior to descent, on November 19, 1969.

completing the mission's objective of a precision landing.

The landing site's lunar coordinates were 3.01239° S latitude and 23.42157° W longitude. The descent had kicked up a high-velocity spray of lunar dust, causing the Surveyor probe to be sandblasted by the fine particles. Interestingly, this process removed more dust from the probe than it deposited. The astronauts observed that portions of Surveyor's surface exposed to the sandblasting had lightened back to their original white color as the layer of lunar dust that had given the probe a tan hue was stripped away.

When Charles "Pete" Conrad, the shortest astronaut of the early NASA crews, stepped onto the lunar surface, he enthusiastically exclaimed, "Whoopie! Man, that may have been a small one for Neil, but that's a long one for me." This remark was not spontaneous. Conrad had made a $500 bet with journalist Oriana Fallaci, claiming he would say those words to prove NASA did not script astronauts in their historic moments. However, Conrad never collected the winnings from the bet.

Alan Bean, the Lunar Module Pilot, was the next to exit the Lunar Module (LM), and their mission proceeded with the deployment of a color television camera, a significant improvement over the monochrome camera used during Apollo 11. Unfortunately, as Bean set up the camera, he inadvertently pointed it at the Sun, which destroyed the camera's Secondary Electron Conduction (SEC) tube. As a result, live television coverage of the Apollo 12 mission ended abruptly.

Charles Conrad Jr. beside the US flag during Apollo 12

After planting the American flag on the Moon, Conrad and Bean dedicated much of their time during the first Extravehicular Activity (EVA) to deploying the Apollo Lunar Surface Experiments Package (ALSEP), designed to collect data from the lunar environment long after their departure. The deployment did not go entirely smoothly—Bean had difficulty removing the plutonium fuel element from its protective cask and had to use a hammer to free it. Despite minor setbacks, they successfully deployed the experiments, including a seismometer capable of detecting their footprints as they moved back to the LM. Before concluding the first EVA, they collected lunar samples, including a core tube full of material from beneath the Moon's surface. The first EVA lasted 3 hours, 56 minutes, and 3 seconds.

Astronaut Alan L. Bean, lunar module pilot for the
Apollo 12 mission, is about to step off the ladder of the
Lunar Module to join astronaut Charles Conrad Jr.,
mission commander,

Four potential geological traverses had been planned based on where the LM might land. Conrad, however, skillfully set the LM down between two planned landing points. During the first EVA and the subsequent rest period, Houston scientists combined two traverses into one circular route for the astronauts to follow. Thirteen hours after the first EVA, Conrad and Bean emerged from the LM to begin their second excursion. Their first stop was Head Crater, located about 100 yards from the LM. There, Bean noticed lighter material beneath Conrad's footprints, suggesting ejecta from Copernicus Crater, 230 miles to the north. This discovery confirmed scientists' hopes and allowed them to date the impact that formed Copernicus to around 810 million years ago.

The astronauts continued their exploration to Bench Crater and Sharp Crater, before reaching Surveyor Crater, where the Surveyor 3 probe had landed two years earlier. Fearing unstable footing or the possibility that the probe might topple, they approached cautiously but found the terrain solid. After examining the probe, they collected several parts, including the television camera, and took samples of rocks that had been studied by the probe's television instruments. The astronauts had hoped to use a camera timer to take a "selfie" with Surveyor, but could not find the timer in their tool carrier. Before heading back to the LM, they visited Block Crater within Surveyor Crater. The second EVA lasted 3 hours, 49 minutes, and 15 seconds, during which they traveled 4,300 feet and collected 73.75 pounds of lunar samples.

Gordon's Solo Lunar Orbit Activities

While Conrad and Bean explored the lunar surface, Command Module Pilot Richard Gordon remained in lunar orbit aboard the Command and Service Module (CSM), Yankee Clipper. Initially, Gordon had little to say as Mission Control focused on the lunar landing, but once Conrad and Bean were safely on the surface, he sent his congratulations. On his next orbit, he spotted both the LM and the Surveyor probe on the ground and conveyed their positions to Houston.

During the first EVA, Gordon prepared for a critical maneuver—a burn to alter the CSM's orbit to account for the Moon's rotation. This plane-change maneuver ensured that Yankee Clipper would be in the correct position to rendezvous with the LM once it launched from the lunar surface. While performing this maneuver, Gordon experienced communication difficulties with Mission Control, as Conrad and Bean were using the same communications circuit. Nevertheless, once the moonwalkers returned to the LM, Gordon executed the plane change successfully.

In addition to navigation duties, Gordon conducted the Lunar Multispectral Photography Experiment. Using four Hasselblad cameras, each fitted with a different color filter, he captured simultaneous images of lunar features at various points in the spectrum. These images provided insights into the composition of potential landing sites for future Apollo missions and revealed details not visible to the naked eye.

Deploying the ALSEP: Apollo 12's Scientific Legacy

The Apollo Lunar Surface Experiments Package (ALSEP) was a suite of scientific instruments designed to be left on the Moon to operate autonomously, transmitting data back to Earth long after the astronauts departed. NASA saw the ALSEP as a way to demonstrate that crewed lunar missions could conduct scientific work more efficiently than robotic spacecraft. The Bendix Corporation was awarded the contract to design and build the ALSEP in 1966. Due to the limited time Apollo 11 astronauts spent on the lunar surface, a smaller set of instruments, known as the Early Apollo Surface Experiment Package (EASEP), had been deployed during that mission. Apollo 12 was the first mission to carry a full ALSEP suite, and each subsequent Apollo mission would carry similar, though sometimes varied, packages.

When the ascent stage took off, Apollo 12's ALSEP was deployed at least 300 feet from the LM to protect the instruments from debris. The package included a Lunar Surface Magnetometer to measure the Moon's magnetic field, a Cold Cathode Gauge to study the density and temperature of the thin lunar atmosphere, and a Solar Wind Spectrometer to analyze the composition of the solar wind. A Passive Seismic Experiment (PSE) was also deployed, intended to measure seismic activity, including moonquakes, and to detect impacts, such as the planned crash of Apollo 12's ascent stage, which would hit the Moon with the force of one ton of TNT.

All the ALSEP instruments were connected to a Central Station, which managed power distribution, communications, and data processing. The station was powered by a SNAP-27 Radioisotope Thermoelectric Generator (RTG) that used plutonium as a fuel source. This was the first use of nuclear power on a crewed NASA mission. The plutonium core had been housed in a protective cask designed to survive re-entry in the event of an aborted mission. This same type of cask would later survive re-entry during the Apollo 13 mission, sinking into the Pacific Ocean's Tonga Trench without leaking any radiation.

Although some instruments encountered difficulties—such as the failure of the Lunar Atmosphere Detector's power supply—the ALSEP package functioned effectively for years. Apollo 12's Passive Seismic Experiment transmitted valuable data about the Moon's interior until NASA deactivated the equipment on September 30, 1977, due to budget constraints.

In all, Apollo 12's scientific experiments and geological samples provided invaluable insights into the Moon's composition, history, and environment, marking another significant achievement in the Apollo program's exploration of the lunar surface.

On November 20, 1969, after a successful mission on the lunar surface, the Intrepid Lunar Module (LM) lifted off from the Moon at 14:25:47 UT (mission time 143:03:47.78). After several carefully calculated maneuvers, the LM docked with the Command and Service Module (CSM), Yankee Clipper, three and a half hours later. Following the docking, at 147:59:31.6, the ascent stage of the LM was jettisoned, and the CSM maneuvered away.

The LM's remaining propellant was depleted in a controlled burn directed by Mission Control, causing Intrepid to impact the lunar surface 39 nautical miles (72 km) from the Apollo 12 landing site. The seismic vibrations generated by the impact were detected by the seismometer the astronauts had left behind on the Moon, with the instrument recording the vibrations for over an hour—an important experiment in understanding the Moon's geophysical properties.

The Apollo 12 crew spent an additional day in lunar orbit, capturing photographs of potential landing sites for future Apollo missions. A second plane change maneuver was executed at 159:04:45.47, lasting 19.25 seconds, positioning the spacecraft for its eventual return to Earth.

The trans-Earth injection burn, the critical maneuver that sent the Yankee Clipper on its trajectory back to Earth, was performed at 172:27:16.81 and lasted 130.32 seconds. Two minor midcourse corrections were made during the return journey. The astronauts also conducted a final television broadcast, during which they answered questions submitted by the media. One of the highlights of the return trip was the crew's observation of a solar eclipse, which occurred when the Earth passed between the spacecraft and the Sun. Lunar Module Pilot Alan Bean described it as the most spectacular sight of the entire mission.

A solar eclipse seen from Apollo 12

Apollo 12's Yankee Clipper splashed down in the Pacific Ocean on November 24, 1969, at 20:58 UT (3:58 pm Eastern Time, 10:58 am HST), completing the mission. The landing was unexpectedly rough, causing a camera to break loose and strike Alan Bean on the forehead. The USS Hornet quickly recovered the crew, and they entered the Mobile Quarantine Facility (MQF), designed to prevent any potential lunar contamination from reaching Earth.

Lunar samples and parts of the Surveyor 3 probe were flown ahead to the Lunar Receiving Laboratory (LRL) in Houston. Meanwhile, the MQF was transported to Hawaii and then flown to Ellington Air Force Base near Houston, arriving on November 29. The Apollo 12 crew remained in quarantine at the LRL until December 10, ensuring that they were free of any possible contaminants from the lunar environment.

After the Mission

Following the mission, Pete Conrad encouraged his fellow astronauts to join him in NASA's Skylab program, seeing it as their best opportunity to return to space. Alan Bean took Conrad's advice and went on to command the Skylab 3 mission, while Conrad himself commanded Skylab 2, the first crewed mission to the space station. Richard Gordon, however, held onto the hope of walking on the Moon and stayed with the Apollo program, serving as the backup commander for Apollo 15. He had been slated to command Apollo 18, but that mission was ultimately canceled, and Gordon never flew in space again.

Meanwhile, after jettisoning the Service Module during re-entry, Mission Control attempted to fire its thrusters to have it skip off the atmosphere into a high-apogee orbit. However, without tracking data to confirm this, it was assumed that the Service Module most likely burned up in the atmosphere along with the Command Module's re-entry.

The ascent stage of Intrepid impacted the Moon on November 20, 1969, at 22:17:17.7 UT (5:17 pm EST) at coordinates 3.94° S latitude and 21.20° W longitude. Years later, in 2009, NASA's Lunar Reconnaissance Orbiter

(LRO) photographed the Apollo 12 landing site, revealing the still-visible descent stage, ALSEP equipment, the Surveyor 3 spacecraft, and the footprints left by the astronauts. In 2011, the LRO returned to the site at a lower altitude, capturing even higher-resolution images of this historic location.

The legacy of Apollo 12 remains preserved on the Moon, a testament to the precision and success of NASA's second manned lunar landing.

Luna 16

First robotic automatic sample return from another celestial body (the Moon)

Luna 16 was an uncrewed Soviet space mission launched in 1970 as part of the ambitious Luna program. This mission marked a significant achievement in the Soviet Union's space exploration efforts, becoming the first robotic probe to land on the Moon and return a sample of lunar soil to Earth. It was the third lunar sample return mission overall, following two successful U.S. Apollo missions. The spacecraft returned 101 grams (3.56 ounces) of lunar soil from the Mare Fecunditatis, a dark, flat plain on the Moon's surface, located in the northeast section of the lunar near side.

The Luna 16 spacecraft was composed of two main stages. The descent stage, a cylindrical body with four landing legs, was equipped with fuel tanks, a landing radar, and a complex of descent engines. It served as the landing platform for the ascent stage. The main descent engine slowed the spacecraft's approach as it neared the lunar surface, with the final descent controlled by a set of lower-thrust jets to ensure a soft landing. The descent stage also housed a television camera, radiation and temperature monitors, and an extendable arm with a drilling rig designed to collect a sample of lunar soil.

Mounted atop the descent stage was the ascent stage, a smaller cylindrical module with a rounded top. This stage contained a hermetically sealed sample container and a re-entry capsule for the soil sample's journey back to Earth. The descent stage remained on the lunar surface after the ascent stage lifted off, transmitting valuable data on lunar temperature and radiation levels.

The Luna 16 mission launched on September 12, 1970, and after a mid-course correction on September 13, the spacecraft entered lunar orbit on September 17. The craft orbited the Moon at an altitude of 111 kilometers, with an inclination of 70 degrees. Over the next two days, a series of orbital adjustments were made to lower its perilune (the closest point to the Moon's surface) to just 15.1 kilometers. On September 20, at 05:12 UTC, the spacecraft's main braking engine fired, initiating its descent toward the lunar surface. Six minutes later, at 05:18 UTC, Luna 16 softly landed at 0°41' south latitude and 56°18' east longitude in the Mare Fecunditatis, approximately 100 kilometers west of Webb Crater and 150 kilometers north of Langrenus Crater.

This landing was historically significant as the first to occur during lunar night, with the Sun having set 60 hours prior. Upon reaching an altitude of 20 meters, the main descent engine cut off, and at 2 meters above the surface, the landing jets ceased operation, allowing the spacecraft to gently free-fall the remaining distance. The total mass of Luna 16 at the time of landing was 1,880 kilograms.

Less than an hour after landing, Luna 16's automated drill began collecting a lunar soil sample, penetrating the surface to a depth of 35 centimeters. The sample was carefully transferred into a small spherical capsule on the spacecraft. After spending 26 hours and 25 minutes on the lunar surface, Luna 16's ascent stage lifted off from the Moon on September 21 at 07:43 UTC, leaving the descent stage behind. Three days later, on September 24, the re-entry capsule containing 101 grams of lunar soil reentered Earth's atmosphere at a velocity of 11 kilometers per second. It successfully parachuted to the ground in Kazakhstan, landing 80 kilometers southeast of the town of Jezkazgan at 05:25 UTC.

The mission was hailed as a major success for the Soviet space program. Luna 16 accomplished the first fully automated recovery of samples from an extraterrestrial body. The basaltic soil samples were similar in composition to those collected by the U.S. Apollo 12 mission. According to observations from the Bochum Observatory in Germany, television images transmitted by the spacecraft were of high quality.

In recognition of the mission's scientific significance, a small portion of the lunar soil, weighing 0.4825 grams, was sent to Britain for further analysis. The Luna 16 mission cemented the Soviet Union's role in lunar exploration, marking a key moment in the ongoing Space Race and the broader effort to explore and understand our closest celestial neighbor.

Lunokhod 1

First lunar rover (remote-controlled)
First rover on another celestial body (the Moon)

Lunokhod 1, known in Russian as Луноход-1 or "Moonwalker 1," was the first robotic rover to traverse the surface of the Moon, marking a major milestone in space exploration. Launched by the Soviet Union as part of the Lunokhod program, Lunokhod 1 was carried to the Moon aboard the Luna 17 spacecraft in 1970. This mission followed a failed earlier attempt with Lunokhod 0 (Device 8EL No. 201), which launched in February 1969 but failed to reach Earth orbit.

Lunokhod 1 was a remarkable engineering achievement. Designed to operate for three lunar days (approximately three Earth months), the rover far exceeded expectations, functioning for eleven lunar days—equivalent to 321 Earth days. Over the course of its mission, Lunokhod 1 traversed a total distance of 10.54 kilometers, leaving a lasting imprint on the history of lunar exploration.

Soviet Lunokhod moonrover Photo credit: Petar
Milošević

Lunokhod 1's structure resembled a tub-like compartment with a large convex lid, standing 135 centimeters (4 feet 5 inches) tall and weighing 840 kilograms (1,850 pounds). The vehicle was 170 centimeters (5 feet 7 inches) long and 160 centimeters (5 feet 3 inches) wide, and it moved across the Moon's surface on eight independently powered wheels. The rover was equipped with two types of antennas: a cone-shaped antenna for general communications and a highly directional helical antenna for precise data transmission.

Its scientific instrumentation included four television cameras and several extendable devices designed to test the lunar soil for density and mechanical properties. The rover also carried an array of scientific tools, including an X-ray spectrometer, an X-ray telescope, cosmic ray detectors, and a laser retro-reflector, the latter supplied by France. The rover's power came from batteries recharged by solar panels mounted beneath the lid. To withstand the harsh lunar environment, the rover's mechanical parts were lubricated with special fluoride-based grease, and its electric motors—one in each wheel—were enclosed in pressurized containers to protect them from the vacuum of space.

During the cold lunar nights, when temperatures plummeted, the rover's lid was closed to conserve heat, and a polonium-210 radioisotope heater unit kept the internal components warm enough to remain operational.

Luna 17 launched on November 10, 1970, at 14:44:01 UTC, with Lunokhod 1 aboard. After reaching Earth orbit, the spacecraft's final stage ignited, sending it on a trajectory toward the Moon. Following two course corrections on November 12 and 14, Luna 17 entered lunar orbit on November 15 at 22:00 UTC.

On November 17, 1970, at 03:47 UTC, Luna 17 successfully soft-landed in the Mare Imbrium (Sea of Rains), one of the Moon's vast, basaltic plains. The landing site was located about 60 kilometers south of Promontorium Heraclides. The lander had dual ramps from which Lunokhod 1 descended onto the lunar surface at 06:28 UTC.

Lunokhod 1's daily operations were dictated by the lunar cycle. The rover worked during the lunar day, pausing to recharge its batteries via the solar panels, and then hibernated during the frigid lunar nights, protected by its radioactive heat source. Along its journey, the rover encountered and named several small craters, such as Leonid, Kolya, Valera, and Borya. These names, given unofficially during the mission, were later officially recognized by the International Astronomical Union (IAU) in 2012.

Lunokhod 1's exploration began immediately after its descent. During its first lunar day (November 17-22, 1970), the rover traveled 197 meters, capturing 14 close-up images of the lunar surface and 12 panoramic views. It also conducted its first soil analysis. In total, during its 322 Earth days of operation, Lunokhod 1 traveled 10.54 kilometers, sent back over 20,000 television images, and produced 206 high-resolution panoramas. It performed 25 soil analyses using its RIFMA X-ray fluorescence spectrometer and probed the lunar soil at 500 different locations.

The rover's detailed mapping of the lunar surface provided invaluable data, helping scientists understand the Moon's geological history. It also demonstrated the feasibility of remote-controlled exploration on another celestial body.

After September 14, 1971, communication with Lunokhod 1 ceased, and all attempts to re-establish contact failed. The mission was officially declared complete on October 4, 1971, coinciding with the anniversary of the launch of Sputnik 1, the first artificial satellite. For decades, the exact location of Lunokhod 1 remained uncertain, until March 17, 2010, when researcher Albert Abdrakhimov discovered both the lander and rover in images taken by NASA's Lunar Reconnaissance Orbiter (LRO).

In April 2010, the Apache Point Observatory Lunar Laser-ranging Operation (APOLLO) team, based at the University of California, San Diego, used these LRO images to pinpoint Lunokhod 1's position accurately. They successfully measured the distance to the rover's laser retro-reflector, which had been dormant for nearly 40 years. Surprisingly, the retro-reflector was still highly functional, returning much stronger signals than other reflectors placed on the Moon. This discovery allowed scientists to use Lunokhod 1 for precise laser ranging experiments, enhancing our understanding of the Earth-Moon system.

By November 2010, the rover's location had been determined to within a centimeter, enabling further studies. The ability to range the rover even when it was in sunlight added to the scientific value of the retro-reflector. In 2013, French scientists replicated the laser-ranging experiments, further solidifying Lunokhod 1's continued contribution to lunar research.

Lunokhod 1's mission was a groundbreaking achievement for robotic space exploration. Despite its projected initially three-month lifespan, the rover exceeded all expectations, paving the way for future lunar and planetary exploration. It provided extensive data about the Moon's surface and soil and demonstrated the viability of long-duration remote-controlled missions on other celestial bodies. Its rediscovery decades later contributes to our understanding of the Moon and the broader Earth-Moon relationship, underscoring the long-term scientific value of this pioneering mission.

Venera 7

1First soft landing on another planet (Venus)

First signals from another planet

Venera 7, part of the Soviet Union's ambitious Venera series of probes, achieved a monumental milestone in space exploration. Launched on August 17, 1970, it became the first spacecraft to successfully soft-land on another planet and transmit data back to Earth from Venus—a feat that forever changed humanity's understanding of our neighboring planet. The mission marked a significant step in the Space Race, as the Soviet Union continued to push the boundaries of planetary exploration.

The Venera 7 spacecraft was meticulously designed to withstand the harsh and unknown conditions of Venus. With the surface of Venus shrouded in thick clouds and extreme heat, the probe was constructed to endure pressures up to 18 megapascals (2,600 psi) and temperatures as high as 580°C (1,076°F). These parameters far exceeded what scientists believed it would encounter, but given the significant uncertainties about Venus's surface environment, the spacecraft's designers built in a large margin of error. The ruggedness of the lander came at a cost, however, as it limited the payload capacity for scientific instruments. Despite this, Venera 7 was equipped with essential tools for atmospheric and surface analysis, including temperature and pressure sensors, an accelerometer to measure atmospheric density, and a radar altimeter. Meanwhile, its interplanetary bus carried detectors for solar wind particles and cosmic rays.

On its journey to Venus, Venera 7 completed two mid-course corrections using its onboard KDU-414 engine. As it neared Venus, the spacecraft descended into its atmosphere on December 15, 1970. In a carefully orchestrated sequence, the lander remained attached to the interplanetary bus during the initial phase of atmospheric entry, allowing the bus to cool the lander to −8°C (18°F), preparing it for the extreme heat that awaited below. When atmospheric turbulence severed the bus's communication lock with Earth, the lander was released.

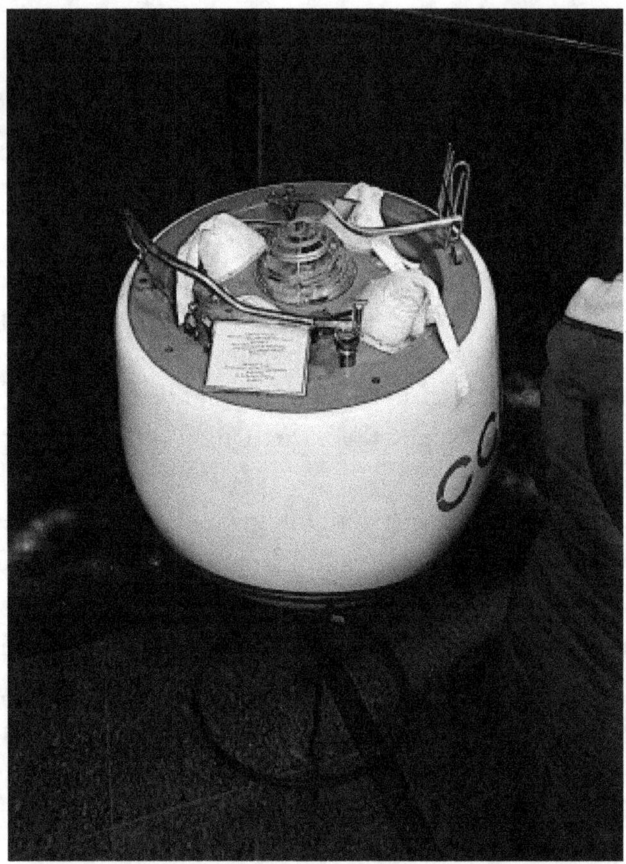

As Venera 7 descended through Venus's dense atmosphere, its parachute opened at 60 kilometers (37 miles) above the surface. However, the descent did not go entirely as planned. After initial success, the parachute began to fail, causing the probe to plummet faster than anticipated. Ultimately, the parachute collapsed completely, and the lander free-fell the remaining distance, striking Venus's surface at approximately 16.5 m/s (59 km/h; 37 mph). The lander impacted near the coordinates of 5°S 351°E, possibly bouncing and coming to rest on its side.

Initial reports suggested that the probe had gone silent upon impact. However, Soviet radio astronomer Oleg Rzhiga's closer review of telemetry tapes revealed faint signals that continued for 23 minutes after landing. Despite the lander's compromised position, it transmitted valuable data for a total of 53 minutes, including approximately 20 minutes from the planet's surface.

The data returned from Venera 7 provided groundbreaking insights into the Venusian environment. The temperature at the landing site was measured at an astonishing 475°C (887°F), and atmospheric pressure was calculated to be around 9 megapascals (1,310 psi), confirming the planet's extreme conditions. These findings dashed any remaining hopes of Venus being a habitable world. The probe's rapid deceleration upon impact further suggested that it had landed on a solid surface, likely with minimal dust, rather than a liquid one.

Venera 7's successful mission not only represented a triumph for Soviet engineering but also delivered definitive proof that Venus's surface was far too hostile for human exploration or the presence of liquid water. The spacecraft's achievement opened a new chapter in interplanetary exploration and underscored the formidable challenges of exploring the inner planets of our solar system.

Salyut 1

First human-crewed space station launched.

Salyut 1, launched by the Soviet Union on April 19, 1971, holds the distinction of being the world's first space station. This groundbreaking achievement marked a significant milestone in the Space Race, as the Soviets shifted their focus toward long-duration orbital missions following the success of Apollo 11. The Salyut program, which spanned several years, would ultimately lead to five more successful launches, cementing the Soviet Union's commitment to space station technology. The final module, Zvezda (DOS-8), became an integral part of the Russian segment of the International Space Station (ISS), continuing its legacy in space exploration.

The design of Salyut 1 was derived from the Almaz military space station program, though it was modified for civilian use. It was composed of five key sections: a transfer compartment, a main compartment, two auxiliary compartments, and the Orion 1 Space Observatory. Salyut 1's purpose was to test the feasibility of prolonged human habitation in space, conduct scientific research, and push the boundaries of technology in a low Earth orbit environment.

Salyut 1 was visited by two spacecraft: Soyuz 10 and Soyuz 11. The Soyuz 10 mission, launched on April 22, 1971, carried cosmonauts Vladimir Shatalov, Aleksei Yeliseyev, and Nikolai Rukavishnikov. Although they successfully soft-docked with the station after a 24-hour approach, a mechanical failure prevented a hard docking, and the crew was unable to enter the station. The mission was aborted, and the cosmonauts safely returned to Earth after 5.5 hours.

Soyuz 11, launched on June 6, 1971, achieved the first successful docking with Salyut 1. The crew—Georgi Dobrovolski, Viktor Patsayev, and Vladislav Volkov—entered the station and remained aboard for 23 days, setting a new record for time spent in space. Their mission included experiments in Earth observation, space navigation, and medical research, as well as the operation of the Orion 1 Space Observatory, which obtained ultraviolet spectrograms of stars.

Tragically, the mission ended in disaster. On June 29, 1971, as Soyuz 11 re-entered Earth's atmosphere, a pressure-relief valve in the spacecraft's reentry capsule malfunctioned, causing the cabin to depressurize. The crew, who were not wearing pressure suits, asphyxiated before landing. This tragedy marked the first—and to date, the only—human deaths to occur beyond the Kármán line, the boundary of space.

Geopolitical pressures and internal Soviet motivations influenced Salyut 1's development. After the success of the American lunar landing, the Soviet space program pivoted toward long-duration space stations to reassert its technological prowess. The failure of the N-1 rocket, intended for a Soviet lunar mission, further shifted focus to orbital stations. The Salyut program was also partially motivated by competition with the United States Skylab, which was in development at the time.

Salyut 1 was internally known as DOS-1, with DOS standing for "long-duration orbital station." Although a civilian program, the station also carried military experiments, including the OD-4 optical visual ranger and the Orion ultraviolet instrument, which analyzed rocket exhaust plumes. The station was launched aboard a Proton rocket and was managed by Kerim Kerimov, chairman of the state commission for Soyuz missions.

Design and Structure of Salyut 1

Salyut 1 was 15.8 meters long, with a maximum diameter of 4.15 meters and a habitable volume of 90 cubic meters. Its launch mass was 18,900 kilograms, supported by four solar arrays that spanned about 10 meters. The interior was divided into several sections, including the main compartment, which housed work consoles, chairs, and observation portholes. The station was equipped with life support systems, communications equipment, and power supplies.

One of the most important innovations aboard Salyut 1 was the SSVP docking system, which allowed internal crew transfer between the Soyuz spacecraft and the station—a precursor to modern docking technologies used on the ISS.

Another significant feature was the Orion 1 Space Observatory, developed by Grigor Gurzadyan of Byurakan Observatory. It allowed for ultraviolet spectrography of stars, a first in space history. Viktor Patsayev, one of the Soyuz 11 crew members, became the first human to operate a telescope outside Earth's atmosphere.

Following the Soyuz 11 tragedy, Soviet engineers redesigned the Soyuz spacecraft to accommodate pressure suits for all future missions. However, the redesign process took longer than anticipated, and by the fall of 1971, Salyut 1 was running low on fuel. The station had also suffered an electrical fire during the Soyuz 11 mission, raising concerns about its long-term viability. On October 11, 1971, after 175 days in orbit, Salyut 1 was deliberately deorbited and burned up upon reentry into Earth's atmosphere over the Pacific Ocean.

Despite its relatively short operational life, Salyut 1 set the foundation for all future Soviet space stations, from the later Salyut modules to the construction of the Mir space station. Its legacy continues through its contributions to long-duration spaceflight, international cooperation, and the advancement of human space exploration.

Salyut 1's mission—although marked by both success and tragedy—played a crucial role in demonstrating the feasibility of human space stations and paved the way for future achievements in low Earth orbit, ultimately leading to the creation of the ISS, a symbol of international collaboration in space.

Soyuz 11 / Salyut 1

First human-crewed orbital observatory (Orion 1)

Soyuz 11 (Russian: Союз 11, lit. 'Union 11') remains one of the most tragic yet significant milestones in space exploration, as it was the only crewed mission to board the world's first space station, Salyut 1. On June 7, 1971, cosmonauts Georgy Dobrovolsky, Vladislav Volkov, and Viktor Patsayev successfully docked with Salyut 1, embarking on a mission that would set endurance records in space but end in disaster. After 22 days in orbit, the mission concluded on June 29, 1971, when the Soyuz 11 capsule depressurized during re-entry preparations, killing the entire crew. To this day, the Soyuz 11 cosmonauts are the only humans known to have died in space.

Cosmonauts Georgy Dobrovolsky, Vladislav Volkov and Viktor Patsayev.

The original crew for Soyuz 11 consisted of Alexei Leonov, Valeri Kubasov, and Pyotr Kolodin. However, just days before launch, Kubasov was suspected of having tuberculosis following a medical X-ray examination, forcing the replacement of the prime crew with their backups. Thus, Georgy Dobrovolsky, commanding his first spaceflight, led flight engineer Vladislav Volkov, on his second spaceflight, and research engineer Viktor Patsayev, also on his first mission. Despite the change in personnel, the mission proceeded as planned, and the crew boarded Soyuz 11.

Soyuz 11 launched from the Baikonur Cosmodrome in Kazakhstan on June 6, 1971, using the callsign Yantar (Amber). Earlier that year, Soyuz 10 had failed to dock with Salyut 1, leaving Soyuz 11 with the responsibility of completing the mission. The docking sequence began as the spacecraft approached Salyut 1, initially controlled by automatic systems before switching to manual control for final adjustments. After several hours, the spacecraft was securely docked to the station on June 7, 1971.

Upon entering the station, the cosmonauts were met with a smoky, burnt atmosphere, requiring them to replace part of the ventilation system and temporarily return to Soyuz 11. Once the air cleared, the crew resumed their activities aboard Salyut 1, marking a period of intense scientific productivity. Their mission included live television broadcasts, various scientific experiments, and testing the effects of long-duration spaceflight on the human body. However, their stay was not without difficulties—on the 11th day, a fire broke out, nearly forcing an evacuation. Mission planners also had to cancel the planned observation of an N1 rocket launch due to delays.

After spending 22 days aboard Salyut 1, the crew prepared to return to Earth. On June 29, 1971, they loaded their scientific specimens and data into Soyuz 11 and undocked from the station. All seemed routine as Soyuz 11 co-orbited Earth and performed its retrofire burn in preparation for re-entry. However, disaster struck during descent. Approximately 12 minutes after the retrofire, explosive bolts separating the descent and service modules malfunctioned, causing the cabin to depressurize instantly.

Soyuz 11 landed in Kazakhstan on the evening of June 29, 1971. When the recovery team opened the capsule, they found the three cosmonauts dead, their bodies still strapped into their seats. Signs of asphyxiation, including dark-blue patches on their faces and blood trailing from their noses and ears, indicated that they had suffocated when the spacecraft's atmosphere was vented into space. The rapid loss of pressure, traced to a faulty ventilation valve, left the crew with less than a minute to react. Viktor Patsayev was found near the valve, suggesting that he may have tried to close it in his final moments.

The Soviet space program immediately launched an investigation into the accident. It was determined that the simultaneous firing of explosive bolts caused the pressure-equalizing valve to open prematurely, exposing the crew to the vacuum of space. Despite being outfitted with biomedical sensors, the cosmonauts were unable to survive the rapid decompression, with cardiac arrest occurring within 40 seconds. The autopsies revealed that the cosmonauts had died from hemorrhaging in the brain and other organs, a result of the extreme pressure drop.

In the aftermath of the tragedy, Soyuz spacecraft were extensively redesigned. The most significant change was the reduction of crew size from three to two, allowing for the inclusion of pressurized Sokol space suits for all cosmonauts during launch and re-entry. This modification aimed to prevent another such fatal depressurization.

The deaths of Dobrovolsky, Volkov, and Patsayev shocked the world, but the Soviet Union highlighted their contributions to space science rather than focusing on the tragic end. The cosmonauts were given a state funeral and interred at the Kremlin Wall Necropolis, a resting place for Soviet heroes. The crew was posthumously awarded the Hero of the Soviet Union medal, and United States astronaut Thomas Stafford served as one of the pallbearers at their funeral.

Memorials to the Soyuz 11 crew can be found worldwide. Craters on the Moon are named after Dobrovolsky, Volkov, and Patsayev, and their names are inscribed on the Fallen Astronaut plaque left on the Moon by Apollo 15 astronauts. In Kazakhstan, near the landing site, a monument was erected in their honor. Though it was vandalized in 2012, Roscosmos restored the site in 2013. Additionally, a group of hills on Pluto was named Soyuz Colles in recognition of the crew's sacrifice.

Despite its tragic end, the Soyuz 11 mission remains a significant moment in the history of space exploration. The lessons learned from the mission have shaped the future of human spaceflight, ensuring the safety of future cosmonauts and astronauts in their quests to push the boundaries of space.

Apollo 15

First human-driven lunar rover, the Lunar Roving Vehicle

In 1962, NASA initiated contracts to construct 15 Saturn V rockets to land a crew on the Moon by 1970. By the time Apollo 11 completed this objective in 1969, nine rockets remained, allowing NASA to plan 10 lunar landings. These remaining missions were intended to delve deeper into lunar science and exploration. A revised spacecraft configuration was developed for the final five missions, with Apollo 15 being the first of these extended "J" missions. The redesigned Lunar Module could support stays of up to 75 hours and carry the Lunar Roving Vehicle to expand surface mobility. The service module was outfitted with a suite of instruments to gather data from lunar orbit.

Originally, Apollo 15 was to land in Censorinus crater as part of a non-extended mission. However, due to budget cuts in 1970, NASA canceled three planned landings, including Apollo 18, 19, and 20. This reshuffling made Apollo 15 the first extended mission, with its landing site moved to Hadley Rille, a region that was initially slated for Apollo 19.

The crew of Apollo 15 included Commander David R. Scott, Command Module Pilot Alfred M. Worden, and Lunar Module Pilot James B. Irwin. Scott, an Air Force officer born in San Antonio, Texas, in 1932, was a veteran astronaut, having flown on Gemini 8 with Neil Armstrong in 1966 and as the command module pilot on Apollo 9 in 1969. Worden, also an Air Force officer, hailed from Jackson, Michigan, and like Scott, attended the U.S. Military Academy at West Point. He earned two master's degrees from the University of Michigan and joined NASA in 1966. James Irwin, born in 1930 in Pittsburgh, Pennsylvania, attended the U.S. Naval Academy and later served in the

Air Force. He earned a master's degree in 1957 and was selected as an astronaut in 1966. Apollo 15 was the only spaceflight for both Worden and Irwin.

Commander, David R. Scott, Command Module pilot, Alfred M. Worden and Lunar Module pilot, James B. Irwin.

The backup crew for Apollo 15 comprised Richard F. Gordon Jr. as commander, Vance D. Brand as command module pilot, and Harrison H. Schmitt as lunar module pilot. Schmitt, a professional geologist, would later fly on Apollo 17. The mission's support crew included Joseph P. Allen, Robert A. Parker, and Karl G. Henize, all scientist-astronauts selected in 1967. These astronauts provided critical support for the mission's scientific objectives, as the prime crew needed assistance with the complex science experiments they were tasked to perform.

Scientific training for Apollo 15 was a high priority. Lee Silver, a geologist from Caltech, played a key role in training the crew, instilling in them a deep appreciation for lunar geology. This emphasis on science directly resulted from growing pressure within NASA to make lunar exploration more scientifically meaningful. The astronauts trained extensively in geology to ensure they could accurately identify and collect valuable lunar samples. The crew also underwent rigorous training with the newly designed Lunar Roving Vehicle, preparing them for the extended exploration they would conduct at Hadley Rille.

Already familiar with the spacecraft as the backup crew for Apollo 12, David Scott, Alfred Worden, and James Irwin were well-prepared to focus much of their training time on geology and lunar sampling techniques as the prime crew for Apollo 15. Scott, determined to maximize the mission's scientific returns, initiated detailed geological training sessions with Caltech geologist Lee Silver in April 1970. This early collaboration allowed Scott and his crew to hone their skills in lunar geology, with Silver playing a crucial role in guiding them through field expeditions and sampling techniques. The addition of Harrison H. Schmitt, a professional geologist, as Apollo 15's backup Lunar Module Pilot (LMP) added a layer of competition and scientific expertise between the prime and backup crews. Schmitt's involvement encouraged a higher level of dedication to scientific objectives.

Backup Apollo 15 Commander Richard Gordon (right)
and backup Lunar Module Pilot Harrison Schmitt during
geology training in Taos, New Mexico. N

The September 1970 cancellation of two Apollo missions elevated Apollo 15 to a J mission, characterized by longer lunar stays and expanded scientific goals, including the use of the Lunar Roving Vehicle (LRV). This transformation was welcomed by Scott, whose approach to the mission reflected not just that of a skilled pilot, but of a true explorer. David West Reynolds, chronicling the Apollo program, described Scott as having "the spirit of a true explorer," determined to push the limits of lunar exploration. The increased mission complexity, including the need for extensive communication to support the experiments and rover activities, required substantial upgrades to infrastructure, including the near-complete rebuilding of the Honeysuckle Creek Tracking Station in Australia.

The crew's geological field trips, conducted monthly over their 20 months of training, prepared them for the challenges of lunar surface exploration. Initially, these trips were traditional geology lessons, with Silver leading the prime and backup commanders and LMPs to sites in Arizona and New Mexico. As the mission's launch date approached, these excursions became more mission-realistic, incorporating mockups of the astronauts' life support backpacks and simulated communication scenarios. Crews practiced describing geological formations to a remote CAPCOM (capsule communicator), who relied on the astronauts' observations to interpret findings. These exercises, conducted in the Rio Grande Gorge and other sites, familiarized the astronauts with communicating landscape details to scientists who could not directly observe the terrain. Scott, taking his geological responsibilities seriously, grew to enjoy the rigorous fieldwork.

In September 1970, NASA's Site Selection Committee narrowed the potential landing sites for Apollo 15 to two: Hadley Rille, a channel on the edge of the Mare Imbrium near the Apennine mountains, and Marius crater, an area of possible volcanic domes. Ultimately, Hadley Rille was chosen, largely due to Scott's preference. The site offered a combination of geological diversity and striking visual appeal, which Scott believed was vital for exploration. Despite the limited high-resolution images of the site—NASA had previously considered Hadley too rough for earlier landings—Scott's enthusiasm for its potential drove the decision. The steep approach trajectory required for the landing, at 26 degrees, was significantly sharper than the 15-degree approaches of previous missions, adding to the challenge. While Scott and Irwin focused heavily on surface geology, Worden's training was centered around the Command and Service Module (CSM), particularly its Scientific Instrument Module (SIM) bay, which would play a key role in gathering orbital data. Worden spent considerable time at North American Rockwell's Downey, California, facility, where the CSM was being constructed.

Commander David Scott takes a photograph during
geology training in Hawaii, December 1970

His geology training, under the guidance of geologist Farouk El-Baz, focused on mapping and identifying lunar features from orbit. Worden learned how to describe these features in a scientifically useful manner, transmitting valuable observations to Earth-based scientists. Often accompanying his crewmates on field trips, Worden participated in these exercises from an aircraft, simulating the speed and perspective he would have while in lunar orbit.

The intense training schedule took a toll on the astronauts' personal lives. Both Worden and Irwin experienced marital difficulties, fearing that the stress could affect their performance on the mission or even jeopardize their roles. Seeking advice from Scott, the two astronauts worried about how their personal issues might reflect on NASA's image of its astronauts. Scott consulted Deke Slayton, Director of Flight Crew Operations, who reassured them that personal issues would not affect their positions as long as they performed their duties. While Irwin and his wife resolved their differences, Worden's marriage ended in divorce prior to the mission.

Apollo 15's spacecraft was carefully designed for the mission's expanded objectives. The Command and Service Module (CSM-112), named Endeavour after HMS Endeavour, carried a tribute to Captain James Cook, whose ship was the first to undertake a purely scientific expedition. This homage reflected the mission's emphasis on science. The Lunar Module (LM-10), named Falcon in honor of the United States Air Force Academy mascot, carried two falcon feathers as a nod to the crew's military service. The spacecraft also included a Launch Escape System and a Spacecraft-Lunar Module Adapter (SLA-19).

The Service Module's SIM bay, housing instruments such as a gamma-ray spectrometer, mapping camera, and mass spectrometer, faced several technical challenges during assembly and testing at the Kennedy Space Center. Many of the instruments arrived late, and testing them on Earth posed difficulties since they were designed to operate in the vacuum of space. Testing the spectrometers' booms, which extended 24 feet, required innovative solutions to mimic space conditions on Earth. Despite these obstacles, the instruments were integrated successfully, although some issues, such as the mass spectrometer boom's occasional failure to retract, persisted during the mission.

The Lunar Module also underwent significant modifications, including enlarged fuel and oxidizer tanks, an extended engine bell for the descent stage, and additional batteries and solar cells for increased power. These upgrades added 4,000 pounds to the module's weight, making it the heaviest lunar module flown in the Apollo program.

If Apollo 15 had flown as an H mission, it would have used different spacecraft: CSM-111 and LM-9. The CSM eventually served in the Apollo-Soyuz Test Project in 1975, while LM-9 remains on display at the Kennedy Space Center Visitor Complex. Endeavour is now displayed at the National Museum of the United States Air Force in Dayton, Ohio, following its transfer from NASA to the Smithsonian in 1974.

Apollo 15's Saturn V rocket (SA-510) was the tenth flight-ready version of the rocket. Given the increased payload for this mission, several adjustments were made to the rocket's configuration and launch trajectory. The launch azimuth was shifted to a more southerly direction, and the parking orbit was lowered to increase the payload capacity by 1,100 pounds. The S-IC stage saw its retrorockets reduced from eight to four, and the outboard engines were burned longer to accommodate the changes. The S-II stage received modifications to reduce pogo oscillations, which had affected previous flights. Despite the complexities of the launch vehicle, preparations went smoothly, even when the rocket was struck by lightning multiple times during its transport to the launch site.

For Apollo 15, astronauts wore redesigned space suits, known as the A7LB. These suits featured improvements that allowed greater mobility, including a waist joint for bending and the ability to sit on the Lunar Roving Vehicle. The suits' liquid cooling systems and life support connections were streamlined for comfort and efficiency. As in previous missions, Scott's suit bore a distinctive red stripe to distinguish him from his crewmates during lunar activities. Worden's suit, though similar to those worn on earlier missions, was optimized for his deep-space EVA, and did not include the additional gear required for surface operations.

The Lunar Roving Vehicle (LRV), a significant technological advancement of the Apollo program, allowed astronauts to travel farther on the lunar surface than ever before. NASA had considered the idea of a vehicle that could operate on the Moon since the early 1960s, envisioning a larger version known as MOLAB, which featured a closed cabin and would have weighed approximately 6,000 pounds (2,700 kg). Early MOLAB prototypes were tested in Arizona, but as the concept of establishing a lunar base seemed increasingly distant, NASA shifted its focus to a lighter, more efficient vehicle. While it was recognized that a rover would enhance the scientific capabilities of the Apollo J missions, the vehicle's mass had to be limited to around 500 pounds (230 kg) to fit within the mission's payload constraints.

NASA's decision to proceed with the development of the Lunar Roving Vehicle came in May 1969, as Apollo 10 returned from its dress rehearsal for the first Moon landing. Boeing was awarded the contract to build three rovers, but the project faced significant cost overruns, particularly with the development of the navigation system. The final cost for the three vehicles totaled $40 million, which drew media scrutiny during a period of declining public interest in space exploration and budget cuts to NASA.

Despite these challenges, the LRV was a marvel of engineering, capable of being folded into a compact space measuring 5 feet by 20 inches (1.5 m by 0.5 m). The rover itself weighed 460 pounds (209 kg), but when carrying two astronauts and their equipment, its total weight reached 1,500 pounds (700 kg). Each of the rover's four wheels was powered by an independent 1/4 horsepower (200 W) electric motor, allowing it to reach speeds of 6 to 8 miles per hour (10 to 12 km/h). This mobility revolutionized lunar exploration, enabling astronauts to travel farther from the Lunar Module and conduct scientific experiments over a wider area.

Apollo 15, launched on July 26, 1971, marked a significant moment in the United States' Apollo program as the ninth crewed mission and the fourth to land on the Moon. This mission, designated as the first of the J-missions, introduced a new era of lunar exploration, emphasizing extended stays and a greater focus on scientific objectives.

It also featured the debut of the Lunar Roving Vehicle (LRV), a significant technological advancement that allowed astronauts to travel farther across the Moon's surface than ever before.

This launch marked the beginning of a critical lunar mission aimed at exploring Hadley Rille, with timing meticulously aligned to ensure proper lighting conditions upon arrival. The launch occurred within the first moments of the available two-hour, 37-minute window, a vital opportunity as a delay past July 27 would have postponed the mission until late August. Commander David Scott, Command Module Pilot Alfred Worden, and Lunar Module Pilot James Irwin were awakened early that morning and, after suiting up and completing final preparations, were transported to Pad 39A, where they boarded their spacecraft approximately three hours before launch. The countdown proceeded flawlessly, and Apollo 15 lifted off without any delays.

Just over 11 minutes after launch, the Saturn V rocket's third stage, the S-IVB, shut down, placing the spacecraft into low Earth orbit. Apollo 15 would remain in this parking orbit for about two hours and 40 minutes, allowing the astronauts and mission control to thoroughly check all systems. After completing 1.5 orbits around Earth, the S-IVB engine was reignited, initiating the trans-lunar injection (TLI) that propelled the spacecraft onto its course toward the Moon.

Following the TLI, the command and service module (CSM), named Endeavour, and the lunar module (LM), named Falcon, were still attached to the now nearly depleted S-IVB stage. As part of the planned procedure, explosive charges separated the CSM from the S-IVB, and Worden expertly maneuvered the CSM to dock with the LM, which was mounted at the end of the booster. Once docking was successful, the combined spacecraft was jettisoned from the S-IVB. The booster, no longer needed, was programmed to impact the Moon's surface, where its crash was detected by seismometers placed by previous Apollo missions, providing valuable data for scientists studying lunar geology. Although the impact occurred 79 nautical miles away from its intended target, the scientific objectives were still achieved.

During the journey to the Moon, the crew encountered a minor but concerning issue—a malfunctioning light on the service propulsion system (SPS), which suggested a possible malfunction with the spacecraft's engine system. After several hours of troubleshooting, the astronauts conducted a test burn to confirm the system's integrity and perform a midcourse correction. Despite the faulty indicator light, the system functioned properly throughout the mission. Post-flight analysis revealed the issue was caused by a tiny piece of wire lodged inside the switch.

As the spacecraft coasted toward the Moon, the astronauts took the opportunity to enter the lunar module for inspection. About 34 hours into the mission, they renewed the LM's atmosphere to remove any contaminants and moved equipment necessary for lunar exploration. During this process, they discovered a broken cover on the tapemeter, a critical instrument used to measure range and approach speed. Despite the break, the ground team determined the instrument would still function correctly, and the crew cleaned up the floating glass debris using a vacuum cleaner and adhesive tape.

Another challenge arose later in the mission when Commander Scott noticed a leak in the water system as he attempted to chlorinate the supply. The source of the leak was difficult to identify, but ground control provided a solution. The astronauts cleaned up the water using towels and hung them to dry in the tunnel connecting the CSM and the LM. Scott humorously remarked that the tunnel resembled a laundry room with the towels hanging out to dry.

On July 29, after several days of smooth sailing, Apollo 15 approached the Moon. A midcourse correction, the second of the mission, was executed with a brief engine burn. Upon reaching the far side of the Moon, the crew initiated the critical lunar orbit insertion (LOI) burn, which had to be conducted out of communication range with Earth. Mission control anxiously awaited confirmation of the burn's success, knowing that if the burn had failed, the spacecraft would reappear from the lunar shadow ahead of schedule. However, when communication resumed, Scott's first comments were not technical, but rather an expression of admiration for the Moon's beauty. While the burn specifics were eventually relayed, Scott's poetic comments delayed the formal report.

The LOI burn lasted 398 seconds and successfully placed Apollo 15 in a stable elliptical lunar orbit, ranging from 170 to 57 nautical miles above the lunar surface. This achievement set the stage for the next phase of the mission, as the astronauts prepared for their descent to the Moon's Hadley-Apennine region, where they would conduct unprecedented scientific exploration.

In preparation for the lunar landing, the Apollo missions refined their approach with each subsequent flight. On the early missions, such as Apollo 11 and 12, the Lunar Module (LM) decoupled from the Command and Service Module (CSM) and was piloted into a lower lunar orbit before beginning the landing attempt. By the time of Apollo 14, NASA implemented a more efficient maneuver to save fuel on the increasingly heavy Lunar Modules. This adjustment, known as Descent Orbit Insertion (DOI), was performed by the Service Propulsion System (SPS) in the CSM while the LM remained docked. This saved fuel and allowed for a smoother descent preparation.

Apollo 15, the first of the "J-missions" which carried enhanced scientific equipment and the Lunar Roving Vehicle (LRV), implemented the DOI procedure with precision. Initially, the spacecraft entered a lunar orbit with its highest point, or apocynthion, directly over the intended landing site at Hadley. A subsequent burn was performed at the opposite side of the orbit, placing Hadley under the spacecraft's lowest point, or pericynthion. This critical DOI burn occurred at 82 hours, 39 minutes, and 49 seconds into the mission, lasting 24.53 seconds. The resulting orbit brought the spacecraft to an apocynthion of 58.5 nautical miles (108.3 kilometers) and a pericynthion of 9.6 nautical miles (17.8 kilometers), a precise setup for the landing.

APOLLO 15 CSM & SIM BAY VIEWED FROM LM DURING RENDEZVOUS.

During the crew's rest period between July 29 and 30, Mission Control observed that mass concentrations on the Moon were causing the orbit to become increasingly elliptical. By the time the crew was awakened, the pericynthion had dropped to 7.6 nautical miles (14.1 kilometers). To correct this and ensure a safer descent, the crew executed a trim maneuver using the Reaction Control System (RCS) thrusters, raising the pericynthion slightly to 8.8 nautical miles (16.3 kilometers) and the apocynthion to 60.2 nautical miles (111.5 kilometers).

With preparations for the landing in full swing, astronauts Dave Scott and Jim Irwin began readying the Lunar Module, named Falcon, for its descent. As they approached the time for undocking from the CSM, a technical issue delayed separation. The crew, along with Houston, determined that the probe instrumentation umbilical was loose. Al Worden, remaining in the CSM, fixed the issue, allowing Falcon to successfully separate at 100 hours, 39 minutes, and 16 seconds into the mission, slightly behind schedule but at an altitude of 5.8 nautical miles (10.7 kilometers). Worden then executed a burn to place the CSM in a stable orbit while Scott and Irwin descended in Falcon toward the Moon's surface.

Powered Descent Initiation (PDI), the crucial burn that would carry the astronauts to the lunar surface, began at 104 hours, 30 minutes, and 9 seconds. Falcon was initially oriented in such a way that the astronauts were on their backs, unable to see the lunar surface. Once the craft performed its planned "pitchover" maneuver, they were positioned upright, with a clear view of the Moon below. However, the landscape did not immediately match the simulations Scott had practiced. An error in their landing path had moved them about 3,000 feet (910 meters) off course, making it difficult for Scott to identify landmarks, including the prominent Hadley Rille.

As they approached the surface, Scott realized they were likely to overshoot the planned landing site. Skillfully adjusting the vehicle's trajectory, Scott maneuvered Falcon toward a smooth area, looking for a suitable landing spot. At an altitude of about 60 feet (18 meters), lunar dust began to obscure their view, kicked up by Falcon's larger-than-usual engine bell, designed to accommodate the heavier load of the Apollo 15 mission. This created a risk of "blowback"—exhaust reflecting off the surface and damaging the engine. The mission planners had impressed upon the astronauts the importance of shutting down the engine the moment contact was made.

Irwin's call of "Contact" came when one of the landing probes touched the lunar surface. Scott immediately cut the engine, allowing Falcon to fall the remaining 1.6 feet (0.49 meters) to the surface. The landing was harder than anticipated, as Falcon dropped at a speed of 6.8 feet per second (2.1 meters per second). This caused Irwin to exclaim "Bam!" as the vehicle settled on the rim of a small crater, tilting slightly at an angle of 6.9 degrees.

Despite the jarring descent, Falcon had safely landed at 104 hours, 42 minutes, and 29 seconds, or 22:16:29 GMT on July 30, 1971, with just 103 seconds of fuel remaining. The spacecraft was about 1,800 feet (550 meters) from the planned landing site. Upon touchdown, Scott reported back to Houston, "Okay, Houston. The Falcon is on the plain at Hadley."

With Falcon set to remain on the lunar surface for nearly three days, Commander David Scott made it a priority to maintain the circadian rhythm the astronauts were accustomed to. As they had landed in the late afternoon, Houston time, the mission plan called for the astronauts to sleep before conducting any lunar surface activities. However, Scott took the opportunity to perform a unique task before their rest. He opened the Lunar Module's top hatch, typically used for docking, and stood for half an hour, observing the surrounding lunar landscape. This activity, known as a stand-up extravehicular activity (EVA), allowed Scott to take photographs and describe the terrain. His training with geologist Lee Silver emphasized the importance of surveying a new field site from a higher vantage point, and the top hatch served that purpose well. Though initial resistance came from mission managers due to concerns about oxygen loss, Scott convinced them of the value of this first-ever stand-up EVA on the Moon. Jim Irwin, the Lunar Module pilot, declined the chance to observe from the hatch, as it would have required detaching his life support system's umbilicals. After repressurizing the cabin, Scott and Irwin removed their spacesuits to sleep, becoming the first astronauts to do so while on the Moon.

During the crew's rest period, Mission Control in Houston monitored a slow but steady oxygen loss. The astronauts were awakened an hour earlier than planned, and the problem was traced to an open valve on the urine transfer device. In his post-mission debriefing, Scott recommended that future crews be awakened immediately under similar circumstances to avoid prolonged exposure to such issues. With the problem resolved, Scott and Irwin donned their spacesuits, depressurized the cabin, and prepared for their first full EVA.

On July 30, 1971, during the Apollo 15 mission, Commander David Scott and Lunar Module Pilot James Irwin became the seventh and eighth humans to walk on the Moon. Their first task was to deploy the Lunar Roving Vehicle (LRV), an essential tool for expanding the scope of their lunar exploration. The LRV, stowed in a folded compartment on the descent stage of the Lunar Module Falcon, presented some challenges during its deployment due to the lander's tilt. Despite this difficulty, and with assistance from Mission Control in Houston, the astronauts successfully unfolded and maneuvered the rover out of its storage bay. A subsequent system check revealed a malfunction in the front-wheel steering; however, the rear-wheel steering remained functional, allowing the rover to operate.

Once operational, Scott became the first person to drive a vehicle on the Moon. His historic words, "Okay. Out of detent; we're moving," marked the start of a new era in lunar exploration. The LRV, equipped with a remotely controlled television camera, allowed scientists on Earth to follow the astronauts' activities in real-time, albeit with less clarity than the high-resolution still images they captured. The rover, a engineering marvel, expanded the astronauts' range, enabling them to reach areas otherwise inaccessible on foot. It carried a plaque that read, "Man's First Wheels on the Moon, Delivered by Falcon, July 30, 1971," symbolizing a significant leap in human ingenuity.

Irwin with the Lunar Roving Vehicle on the Moon. Mons
Hadley is in the background.

The Apollo 15 mission marked the first use of the LRV, which played a crucial role in exploring the Moon's geological features. Scott and Irwin drove the rover to various sites of scientific interest, significantly contributing to the mission's objectives. Pre-launch tests on Earth had included adding extra structural bracing to ensure the vehicle could withstand the weight of astronauts in Earth's gravity, a precaution unnecessary in the Moon's lower gravity environment.

In addition to surface exploration, Apollo 15 also marked a new phase in lunar scientific research. The crew deployed the Particles and Fields Subsatellite (PFS-1) from the Command and Service Module's Scientific Instrument Module bay. This small satellite, launched into lunar orbit shortly before the crew left the Moon, was designed to study the plasma, particle, and magnetic field environment around the Moon. Its instruments provided data crucial for understanding the Moon's gravitational anomalies, known as mass concentrations (mascons), which had been discovered during earlier Apollo missions. PFS-1 transmitted valuable data from August 1971 until January 1973, when it ceased communication. Although the satellite likely crashed into the Moon's surface after its mission ended, it left behind a wealth of information that advanced scientific understanding of the lunar environment.

One of the primary geological objectives for Scott and Irwin was to explore Elbow Crater, situated along Hadley Rille. As they approached the crater, the striking view of the rille—a deep, sinuous canyon—provided a dramatic backdrop for their work. Elbow Crater was a significant location, as it allowed scientists to pinpoint the lander's position on the Moon accurately. The astronauts collected rock and soil samples from the site before driving the rover to the slopes of Mons Hadley Delta, where they gathered additional material. These samples, along with the deployment of the Apollo Lunar Surface Experiments Package (ALSEP), formed the core of the scientific work during the mission's first extravehicular activity (EVA), which lasted six hours and 32 minutes.

The second EVA took place on August 1, with the LRV's front steering now fully functional. Scott and Irwin drove to the Apennine Front, a geological feature at the base of Mons Hadley Delta, where they conducted extensive sampling of rocks and craters. One of their most notable discoveries occurred at Spur Crater, where they found the

Genesis Rock, an anorthosite believed to be a piece of the Moon's primordial crust. This discovery was a highlight of the mission, fulfilling a key scientific objective. After returning to the Lunar Module, Scott resumed drilling for the heat flow experiment, a task he had struggled with during the first EVA. This time, he completed the drilling successfully. The second EVA lasted seven hours and 12 minutes, during which the astronauts also raised the U.S. flag and conducted soil mechanics experiments.

The third and final EVA focused on retrieving a core sample from the lunar surface. This task, which had been delayed from the previous day, took considerable effort. Despite the difficulties, Scott and Irwin managed to extract the sample, which provided valuable insights into the Moon's geological history. Time constraints forced the astronauts to abandon plans to visit the North Complex, a secondary geological target, and instead, they returned to Hadley Rille for further exploration. Near the end of the EVA, Scott performed a simple yet profound demonstration inspired by Galileo's theory of gravity. Using a falcon feather and a hammer, he dropped both objects simultaneously. In the Moon's low-gravity, airless environment, the feather and hammer hit the surface at the same time, visually confirming Galileo's principle that objects fall at the same rate in the absence of air resistance.

Before departing the lunar surface, Scott left behind a small aluminum statuette called "Fallen Astronaut," along with a plaque listing the names of 14 American astronauts and Soviet cosmonauts who had perished in the pursuit of space exploration. He also placed a Bible on the rover's control panel before leaving the vehicle for the last time. The third EVA lasted 4 hours, 49 minutes, bringing the total time spent outside the Lunar Module to 18.5 hours. Scott and Irwin collected approximately 170 pounds (77 kilograms) of lunar samples, marking Apollo 15 as one of the most scientifically productive missions of the Apollo program.

The Genesis Rock

After the departure of the Lunar Module Falcon, Al Worden, aboard the Command and Service Module (CSM) Endeavour, executed a burn to move the spacecraft into a higher lunar orbit. While Falcon was on the Moon, the mission split into two distinct operations: Worden and the CSM managed by their own Capsule Communicator (CAPCOM) and flight support team, while David Scott and James Irwin focused on lunar surface activities.

Worden's responsibilities while in solo flight centered on photographing the Moon and operating the scientific instruments within the Scientific Instrument Module (SIM) bay of the service module. During the translunar coast, the door to the SIM bay had been explosively jettisoned, revealing an array of instruments designed to study the Moon and its environment. The SIM bay housed several sophisticated instruments, including a gamma-ray spectrometer, an X-ray spectrometer, and a laser altimeter, which unfortunately failed midway through the mission. Two cameras—a stellar camera and a metric camera—comprised the mapping camera system, supported by a panoramic camera derived from reconnaissance technology. These cameras, along with the altimeter, enabled precise documentation of the Moon's surface, identifying the exact time and location of each image captured.

Additional instruments in the SIM bay included an alpha particle spectrometer, which was designed to detect signs of lunar volcanic activity, and a mass spectrometer mounted on a boom, intended to avoid contamination from the spacecraft itself. However, this boom presented challenges, as Worden occasionally had difficulty retracting it after use.

As Endeavour orbited the Moon, Worden was scheduled to pass over the Falcon's landing site, but he was unable to spot the Lunar Module on his initial orbit. It wasn't until a subsequent orbit that he finally observed the landing site. Meanwhile, Houston kept Worden updated on Scott and Irwin's activities on the lunar surface. Despite some minor issues, such as the malfunctioning panoramic camera and a non-functional mission timer, Worden diligently captured numerous images and observations that would later inform the decision to send Apollo 17 to the Taurus-Littrow region in search of volcanic evidence. He even engaged in a lighthearted tradition by greeting Earth after each pass over the far side of the Moon in various languages—a gesture devised in collaboration with his geology instructor, Farouk El-Baz.

The scientific results gathered from Worden's operations in lunar orbit were significant. The X-ray spectrometer detected greater fluorescent X-ray flux than expected, revealing that the lunar highlands were richer in aluminum compared to the maria, the vast dark plains formed by ancient volcanic eruptions. From his unique vantage point, Worden also observed previously unknown lunar features, supplementing his photographs with detailed descriptions.

As Scott and Irwin prepared to lift off from the Moon to return to Endeavour, the CSM's orbit had drifted slightly due to the Moon's rotation, necessitating a plane change burn. Worden successfully executed an 18-second burn with the Service Propulsion System (SPS) to align the CSM's orbit with that of the Lunar Module (LM).

On August 2, 1971, Falcon lifted off from the Moon at 17:11 GMT after 66 hours and 55 minutes on the lunar surface. The docking with Endeavour occurred just under two hours later. Following the transfer of samples and other items from the LM to the CSM, the LM was jettisoned and intentionally crashed into the lunar surface, where its impact was recorded by seismometers left behind by previous Apollo missions. The jettison process was delayed due to difficulties with airtight seals, but once complete, the crew resumed their duties. Despite earlier concerns about irregular heart rhythms in both Scott and Irwin, the crew was not informed during the mission. NASA doctors later theorized that the irregularities were due to potassium deficiencies from their strenuous work on the lunar surface.

The crew spent the next two days conducting orbital science experiments and releasing a subsatellite. On August 4, Endeavour departed lunar orbit with a 2-minute and 21-second burn of the SPS. The following day, Worden performed a historic 39-minute extravehicular activity (EVA) in deep space, retrieving film cassettes from the SIM bay with assistance from Irwin, who remained at the CSM's hatch. At 171,000 nautical miles from Earth, this marked the first-ever "deep space" EVA, a feat only repeated twice more in history, both during subsequent Apollo missions.

On August 7, as the spacecraft neared Earth, the service module was jettisoned, and Endeavour re-entered the Earth's atmosphere. Despite one of the three parachutes failing during descent—likely damaged by fuel venting—the two remaining parachutes ensured a safe landing in the North Pacific Ocean. The crew, after completing a mission lasting 12 days, 7 hours, 11 minutes, and 53 seconds, was recovered by the USS Okinawa and returned home, having set new records and further expanded humanity's understanding of the Moon.

The primary mission objectives of Apollo 15 were ambitious and focused on advancing lunar exploration in the Hadley–Apennine region. These goals included conducting geological inspections, surveys, and sampling of lunar materials, as well as emplacing and activating surface experiments. The mission also aimed to evaluate the extended stay capability of Apollo equipment on the lunar surface, facilitate increased extravehicular activities (EVAs), and improve surface mobility through the use of the Lunar Roving Vehicle (LRV). Additionally, inflight experiments and photographic tasks from lunar orbit were included in the objectives.

Apollo 15 successfully met all these objectives. The astronauts collected a wealth of scientific data and samples, and the mission achieved significant milestones, such as the first use of the LRV on the Moon. However, one photographic objective—to capture images of the gegenschein (a faint light effect opposite the Sun)—was not completed due to the camera not being correctly positioned.

According to the Apollo 15 Mission Report, the mission was the fourth lunar landing and resulted in the acquisition of substantial scientific information. The Apollo system demonstrated its capabilities not only as a means of transportation but also as an effective operational scientific facility. Public interest in the mission was bolstered by the presence of the LRV, which captured audiences' fascination, along with the striking Hadley Rille site and enhanced television coverage.

As David Woods noted in the Apollo Lunar Flight Journal, while later missions traveled further and collected more samples, Apollo 15 stands out as a major human achievement. The combination of advanced science, technology, and exploration at Hadley Base—set against a backdrop of massive mountains and a winding rille—created a moment of awe and admiration for human ingenuity.

Despite the mission's technical and scientific success, the crew members' careers were overshadowed by several controversies. One significant issue arose from a deal made by the crew to carry postal covers to the Moon. David Scott, James Irwin, and Al Worden had agreed to transport approximately 400 postal covers in exchange for a payment of $7,000 each, which they intended to set aside for their children. Walter Eiermann, a well-connected intermediary with NASA staff, facilitated the deal with German stamp dealer Hermann Sieger. The postal covers were carried aboard the Lunar Module Falcon and remained on the lunar surface during the crew's activities. After returning to Earth, 100 of the covers were given to Sieger, who began selling them in late 1971 at $1,500 each.

However, the astronauts had not sought the required permission from Deke Slayton, the head of the astronaut office. When the unauthorized covers came to light in 1972, the crew was removed as the backup team for Apollo 17 and reprimanded for poor judgment. Although the astronauts returned the payments and accepted no compensation, the controversy tarnished their reputations, and none of the three flew in space again. After Worden filed a lawsuit, the covers still in their possession were returned in 1983, a development viewed by some as a form of exoneration.

Another dispute involved the Fallen Astronaut statuette that Scott left on the Moon as a memorial to honor astronauts and cosmonauts who had died. The statuette, sculpted by Belgian artist Paul Van Hoeydonck, was intended to be a simple tribute, with minimal publicity and no commercial reproductions. However, Van Hoeydonck later claimed that he had been under the impression that he would receive recognition as the artist and be permitted to sell replicas. After pressure from NASA, Van Hoeydonck canceled plans to sell 950 signed copies. In 2021, Scott published a document asserting that NASA personnel had designed and fabricated the statuette, although his earlier testimony before a Senate committee had indicated that Van Hoeydonck had created it at his request.

During congressional hearings, a further controversy emerged concerning two Bulova timepieces that Scott had taken on the mission without informing Slayton. Bulova, which had sought to have its watches used on Apollo missions but was passed over in favor of Omega, provided Scott with a prototype watch that he wore during his third EVA after the crystal on his NASA-issued Omega Speedmaster had popped off. Images of Scott wearing the Bulova watch while saluting the American flag on the Moon added to its significance. In 2015, this watch sold for $1.625 million, making it one of the most valuable astronaut-owned artifacts ever auctioned.

Mariner 9

First spacecraft to orbit Mars

Mariner 9, part of NASA's Mariner program, became a milestone in space exploration as the first spacecraft to orbit another planet. Launched on May 30, 1971, from Cape Canaveral, Florida, aboard an Atlas-Centaur rocket, Mariner 9's mission was to explore and map Mars, surpassing its predecessors in scope and detail. The spacecraft reached Mars on November 14, 1971, narrowly ahead of the Soviet Union's Mars 2 and Mars 3 probes, both of which arrived just weeks later. However, a massive planet-wide dust storm greeted Mariner 9, obscuring the Martian surface and delaying its primary imaging mission.

Mariner 9's objectives were to continue atmospheric studies begun by Mariners 6 and 7, mapping over 70% of the Martian surface from an altitude of 1,500 kilometers, with image resolutions ranging from 1 kilometer to 100 meters per pixel. This mission also included the study of temporal changes in Mars' atmosphere and surface, the analysis of the planet's two moons, Phobos and Deimos, and the search for volcanic activity using an infrared radiometer. Originally, NASA had planned a dual-probe mission, similar to Mariners 6 and 7, but the failure of Mariner 8 in its launch on May 8, 1971, forced NASA to rely solely on Mariner 9 for the 1971 Mars launch window.

After its arrival, Mariner 9 faced several months of dust storms so severe that its mission had to be delayed. The orbiter's computer was reprogrammed from Earth, allowing it to wait for the dust to settle. When the storm subsided in mid-January 1972, Mariner 9 began its historic imaging of the Martian surface. Over the course of its mission, it sent back 7,329 images, covering 85% of Mars. These images unveiled unprecedented details about the planet, including massive extinct volcanoes such as Olympus Mons—the largest volcano in the solar system—and the vast canyon system of Valles Marineris, a feature so significant that it was later named in honor of Mariner 9.

The mission also provided evidence of ancient riverbeds, weather fronts, fog, wind and water erosion, and craters—revealing Mars as a world far more dynamic than previously imagined. Mariner 9 was also the first mission to photograph Mars' moons, Phobos and Deimos, adding further depth to its contributions to planetary science. Its findings laid the groundwork for future missions, particularly the Viking program, which would later explore the planet in even greater detail.

The spacecraft was equipped with a suite of scientific instruments designed to maximize its observational capabilities. These included an ultraviolet spectrometer, an infrared interferometer spectrometer, and a visual imaging system that far surpassed the capabilities of earlier Mars probes, achieving a resolution of 98 meters per pixel. The spacecraft's propulsion system allowed for precise maneuvers in Martian orbit, with an RS-2101a engine capable of five restarts, fueled by monomethyl hydrazine and nitrogen tetroxide. A sophisticated attitude control system, featuring nitrogen jets, gyroscopes, and star trackers, ensured that the spacecraft could be oriented accurately for its imaging and scientific tasks.

One of the most notable technological innovations of the Mariner 9 mission was the use of forward error-correcting codes (FEC) to ensure the integrity of the data transmitted back to Earth. The spacecraft's images, transmitted in grayscale, were encoded with a Hadamard code, a type of error-correcting code that allowed for the reconstruction of most image data, even in the face of significant noise interference. This system, coupled with the spacecraft's cutting-edge electronics, represented a major step forward in space communications technology.

Mariner 9's achievements were numerous, not only in terms of its scientific findings but also in demonstrating the effectiveness of orbiting missions over flyby missions. Unlike the Soviet Mars 2 and Mars 3 probes, which struggled to gather data due to the unexpected dust storm, Mariner 9's adaptability allowed it to complete its objectives despite the initial challenges successfully. After nearly a year in orbit, the spacecraft depleted its supply of attitude control gas and was deactivated on October 27, 1972, having left an indelible mark on our understanding of Mars.

As of the latest reports, Mariner 9's precise fate remains uncertain. Initially, NASA estimated that the spacecraft would remain in orbit around Mars for at least 50 years, with the possibility that it might re-enter the Martian atmosphere and burn up or impact the surface by 2022. However, updates on its status suggest that it may have already entered the atmosphere and either burned up or crashed into Mars by 2023.

Mars 2

First hard landing on Mars

The Mars 2 mission was part of the Soviet Union's ambitious Mars program, a series of uncrewed missions aimed at exploring the Red Planet. Launched on May 19, 1971, Mars 2 marked a significant milestone as the first human-made object to reach the surface of Mars, despite the unfortunate failure of its landing system. The mission consisted of both an orbiter and a lander, with the orbiter being modeled after the successful Venera 9 bus and designated as type 4MV. Both the Mars 2 and Mars 3 missions were identical in design and utilized the same Proton-K heavy launch vehicle with a Blok D upper stage for propulsion.

The Mars 2 spacecraft was launched from Baikonur Cosmodrome on May 19, 1971, at 16:22:44 UTC. The spacecraft, weighing a total of 4,650 kilograms (10,250 pounds), consisted of a 3,440-kilogram (7,580-pound) orbiter and a 1,210-kilogram (2,670-pound) lander. After the first stage of the Proton-K rocket separated, the second stage ignited, propelling Mars 2 into a parking orbit around Earth. The Blok D upper stage then executed a final burn to place the spacecraft on its trans-Mars trajectory.

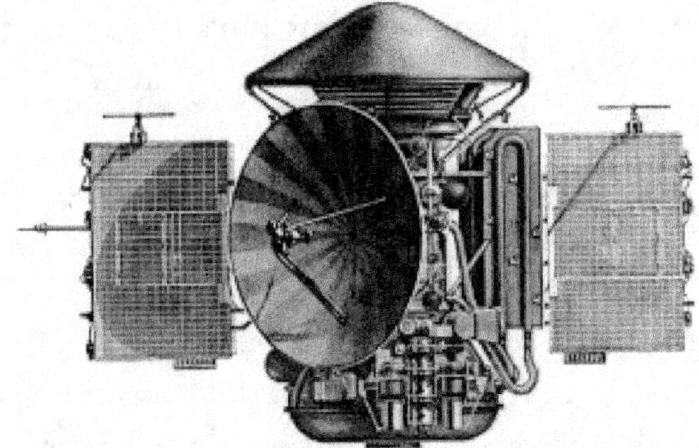

The Mars 2 orbiter successfully entered Mars orbit on November 27, 1971, after a long journey. The spacecraft performed an engine burn to establish an elliptical orbit around the planet, ranging between 1,380 and 2,494 kilometers above the Martian surface, with an inclination of 48.9 degrees. The orbiter's primary mission was to capture detailed images of the Martian surface and clouds, study the planet's atmospheric composition, and measure various environmental parameters, including temperature, pressure, and solar wind activity.

Unfortunately, a massive dust storm enveloped Mars in 1971, complicating the mission's objectives. When NASA's Mariner 9 arrived on November 14, 1971, Mars was shrouded in thick dust, obscuring the surface entirely. Despite this, the Mars 2 and Mars 3 orbiters gathered valuable data during their missions. The dust storm impacted the orbiters' ability to image the surface, but they managed to send back crucial scientific data, including the discovery of surface temperatures ranging from -110°C to 13°C (-166°F to 55°F) and water vapor concentrations 5,000 times lower than Earth's atmosphere.

Mars 2 continued to transmit data until August 1972, completing 362 orbits. Both Mars 2 and its twin, Mars 3, provided valuable insights into Mars' gravity, atmospheric composition, and magnetic fields, and enabled the creation of the first surface relief maps of the planet.

The Mars 2 lander was designed to descend to the Martian surface and conduct scientific experiments. Housed within a conical braking shield, the 1.2-meter diameter spherical landing capsule contained scientific instruments, including two television cameras, a mass spectrometer, temperature and pressure sensors, and a mechanical scoop designed to search for organic materials. The lander was powered by batteries that were charged by the orbiter before separation and sterilized to avoid contaminating the Martian environment.

One of the mission's most intriguing aspects was the inclusion of the small PrOP-M rover, a 4.5-kilogram (9.9-pound) vehicle designed to traverse the Martian surface using skis. The rover was equipped with instruments to measure soil density and radiation levels and was connected to the lander by a 15-meter umbilical cord. Unfortunately, due to the lander's failure, the rover never had the chance to explore Mars.

On November 27, 1971, the descent module separated from the orbiter as planned. However, something went wrong during entry into the Martian atmosphere. The descent system malfunctioned, possibly due to an incorrect angle of entry, and the parachute system failed to deploy. Mars 2 became the first human-made object to crash into the surface of Mars, with its impact site estimated at 45°S 313°W. Despite attempts, contact with the lander could not be established after the crash.

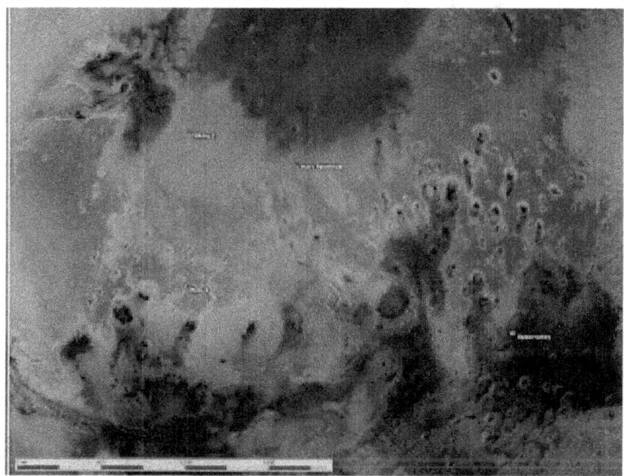

Though the lander was lost, the Mars 2 mission provided the Soviet Union—and the world—with critical data about Mars. The mission's orbiter offered a clearer understanding of the planet's atmosphere, magnetic field, and surface conditions, which would inform future Mars exploration. Along with its twin Mars 3, which succeeded in soft-landing on Mars but also failed shortly after touchdown, the Mars 2 mission represented a bold step in humanity's quest to explore the cosmos.

Mars 3

First soft Mars landing
First signals from Mars surface

Mars 3 was a robotic space probe launched as part of the Soviet Union's Mars program, which aimed to explore the Martian surface and atmosphere. Following the launch of its twin spacecraft, Mars 2, just nine days earlier, Mars 3 was sent on its journey to Mars on May 28, 1971, aboard a Proton-K rocket with a Blok D upper stage. Both Mars 2 and

Mars 3 were nearly identical spacecraft, consisting of an orbiter and a lander, designed to conduct in-depth scientific studies of Mars and its environment. Mars 3 made history as the first spacecraft to achieve a soft landing on Mars, on December 2, 1971, though its success was short-lived, as it ceased transmission just seconds after landing.

Orbiter Mission and Objectives

The primary function of the Mars 3 orbiter, classified as a 4MV-type spacecraft, was to study Martian topography, surface composition, and atmospheric conditions. It also monitored solar radiation, the solar wind, and both the interplanetary and Martian magnetic fields. Additionally, the orbiter served as a vital communications relay between Earth and the lander, transmitting signals and data during the mission.

The mission, however, faced significant challenges from the start. Upon entering orbit around Mars, the orbiter encountered fuel issues, resulting in an incomplete orbital insertion burn. Instead of the planned 25-hour orbit, the spacecraft was placed into a much more elongated, 12-day and 19-hour elliptical orbit. To make matters more difficult, the Martian atmosphere was engulfed in a massive dust storm, the largest ever recorded at the time. The storm, which had already hindered NASA's Mariner 9 mission two weeks earlier, obstructed the surface of the planet, leaving both Mars 2 and Mars 3 orbiters with little choice but to capture images of thick dust clouds rather than the Martian landscape.

Despite these difficulties, the Mars 3 orbiter successfully transmitted data back to Earth from December 1971 through August 1972. Throughout its mission, the orbiter completed 20 orbits and collected crucial data about the Martian environment, including surface temperatures that ranged from −110 °C to +13 °C, surface pressures between 5.5 and 6 millibars, and water vapor concentrations 5,000 times lower than Earth's atmosphere. The orbiter also revealed the existence of atomic hydrogen and oxygen in Mars' upper atmosphere and provided insights into the planet's gravity and magnetic fields. Alongside Mars 2, Mars 3 sent back a total of 60 images, contributing to the creation of surface relief maps and furthering the understanding of Martian topography.

Mars 3's lander was a sophisticated engineering feat, featuring a spherical landing capsule with a diameter of 1.2 meters. This capsule was housed within a 2.9-meter conical aerodynamic braking shield designed to protect the lander during its descent through the Martian atmosphere. The descent system was equipped with main and auxiliary parachutes, retro-rockets, and a radar altimeter, all designed to ensure a controlled landing.

Once the lander reached the surface, four triangular petals were to open, stabilizing the craft and exposing its scientific instruments. The lander carried two television cameras with a 360-degree field of view, sensors to measure temperature, pressure, and wind speed, as well as a mechanical scoop intended to search for organic materials. The lander also carried a pennant featuring the Soviet coat of arms, symbolizing the USSR's achievement.

A key component of the mission was the PrOP-M rover, a 4.5-kilogram, ski-equipped robotic vehicle designed to traverse the Martian surface while tethered to the lander by a 15-meter umbilical cord. The rover was equipped with instruments to measure the mechanical properties of the Martian soil, as well as a radiation densitometer. Although it was intended to operate autonomously, avoiding obstacles with the help of metal rods, the rover never got the chance to explore the surface due to the lander's unexpected failure.

Soviet mars rover PrOP-M.

On December 2, 1971, at 09:14 UTC, Mars 3 released its descent module approximately 4 hours and 35 minutes before reaching Mars. The module entered the Martian atmosphere at a velocity of approximately 5.7 kilometers per second. After enduring aerodynamic braking, parachutes, and retro-rocket deployment, the lander successfully touched down at 45°S 202°E on the Martian surface, becoming the first spacecraft to achieve a soft landing on the Red Planet.

However, the success was short-lived. Ninety seconds after landing, the Mars 3 lander began transmitting a signal to the orbiter, but only 20 seconds later, all communications ceased. The cause of the failure remains unclear, but it is believed that the intense dust storm on Mars at the time may have triggered a coronal discharge, damaging the lander's communications system. The only data received was a partial image—70 lines of gray, with no discernible details.

Despite the abrupt end of its lander operations, the Mars 3 mission contributed significantly to the exploration of Mars. The combined efforts of the Mars 2 and Mars 3 orbiters provided a wealth of data about the Martian environment, contributing to humanity's understanding of the planet's atmospheric and surface conditions. The mission remains a pioneering achievement, representing one of the Soviet Union's most notable efforts during the Space Race.

In 2013, over four decades after the Mars 3 mission, NASA's Mars Reconnaissance Orbiter (MRO) may have imaged the Mars 3 lander hardware on the Martian surface. The HiRISE camera aboard the MRO captured images that showed what might be the lander's parachute, heat shield, and retro-rockets. Amateur space enthusiasts combed through publicly available images to make this discovery, reigniting interest in the legacy of the Mars 3 mission.

Though its lander failed shortly after touchdown, Mars 3's achievement as the first spacecraft to land on Mars remains a monumental milestone in space exploration history.

Pioneer 10

First spacecraft sent on escape trajectory away from the Sun
First mission to enter the asteroid belt and leave inner Solar System
First Jupiter flyby
First Mercury flyby

Pioneer 10, originally designated as Pioneer F, was a groundbreaking NASA space probe launched on March 3, 1972, at 01:49:00 UTC by an Atlas-Centaur rocket from Cape Canaveral, Florida. This mission marked the first successful journey to Jupiter and the first spacecraft to achieve the escape velocity necessary to leave the Solar System. Pioneer 10 became one of the few artificial objects, and the first of five planetary probes, to exit the solar realm. The project was conducted by NASA's Ames Research Center in California, with the spacecraft manufactured by TRW Inc.

The spacecraft was a marvel of engineering for its time, built around a hexagonal bus and equipped with a 2.74-meter parabolic dish antenna for high-gain communication. Pioneer 10's electric power was supplied by four radioisotope thermoelectric generators, which provided a combined 155 watts of power at launch. The probe's design included spin stabilization around the axis of the antenna, ensuring steady orientation during its mission.

On its journey, Pioneer 10 became the first spacecraft to traverse the asteroid belt, completing this feat between July 15, 1972, and February 15, 1973. As it approached Jupiter, Pioneer 10 began photographing the giant planet on November 6, 1973, from 25 million kilometers. During its flyby, which reached its closest approach on December 3, 1973, at a range of 132,252 kilometers, the spacecraft transmitted approximately 500 images. Pioneer 10's suite of instruments allowed it to study Jupiter's atmosphere, magnetic fields, radiation belts, and moons. It also collected valuable data on the solar wind, cosmic rays, and the asteroid belt—insight that would prove crucial for future interplanetary missions.

The Pioneer program itself, part of NASA's broader effort to explore the outer planets, had its roots in the 1960s. American aerospace engineer Gary Flandro of the NASA Jet Propulsion Laboratory conceived the idea of the "Planetary Grand Tour" to take advantage of a rare alignment of the outer planets. Although the Voyager probes would later realize the grand tour in the late 1970s, Pioneer 10 and its twin, Pioneer 11, were launched as preparatory missions.

In February 1969, NASA approved the construction of two spacecraft—Pioneer 10 and 11—to probe beyond Mars and into the unknown regions of the outer Solar System. Both spacecraft were intended to traverse the asteroid belt and explore Jupiter's environment. The scientific rationale for these missions was laid out by the Outer Space Panel, chaired by the renowned American space scientist James A. Van Allen. The mission's objectives were ambitious: explore the interplanetary medium, study the asteroid belt's potential hazards to spacecraft, and gather data on Jupiter's atmosphere and radiation belts.

NASA Ames was selected to manage the Pioneer project due to its expertise with spin-stabilized spacecraft, and TRW Inc. was awarded the contract to build both Pioneer 10 and 11. The design and construction required approximately 25 million man-hours, as the spacecraft needed to be small, lightweight, and capable of surviving the harsh conditions of deep space. Pioneer 10 had a launch mass of about 260 kilograms and carried 36 kilograms of hydrazine monopropellant for orientation control. Six thrusters allowed the spacecraft to maintain a constant spin rate, control forward thrust, and adjust its attitude as it scanned Earth in its orbit. The spacecraft's design also included thermal control mechanisms to ensure the onboard instruments operated within their required temperature ranges.

Pioneer 10's mission provided key data on the radiation environment around Jupiter, even at the risk of damaging some of its systems. The spacecraft's flyby trajectory was chosen to maximize scientific return, allowing scientists to observe the planet's radiation belts in unprecedented detail. After its successful encounter with Jupiter, Pioneer 10 continued on its trajectory towards the outer Solar System. By the time NASA lost radio contact with Pioneer 10 on January 23, 2003, due to a loss of power for its radio transmitter, the spacecraft was 12 billion kilometers from Earth—more than 80 astronomical units (AU) away.

Pioneer 10 was powered by four SNAP-19 radioisotope thermoelectric generators (RTGs), which were positioned on two trusses, each measuring 3 meters in length and spaced 120 degrees apart. This arrangement ensured the RTGs were at a safe distance from the sensitive scientific instruments onboard. At launch, the RTGs provided 155 watts of power, which decayed to 140 watts during its journey to Jupiter. The spacecraft required approximately 100 watts to power all systems. The RTGs were powered by plutonium-238, a radioisotope with a half-life of 87.74 years, encased in a multi-layered capsule protected by a graphite heat shield.

Originally designed to provide power for two years in space, the SNAP-19 RTGs far exceeded expectations. However, despite the longevity of the plutonium-238 fuel, the thermocouple junctions, responsible for converting heat into electricity, deteriorated over time. By 2001, Pioneer 10's power output had diminished to 65 watts, forcing mission controllers to operate instruments selectively.

A sophisticated system of transceivers facilitated communications with Earth. A primary transceiver was linked to the spacecraft's high-gain antenna, a 2.74-meter parabolic dish made from an aluminum honeycomb material. Pioneer 10 was spin-stabilized, with its spin axis aligned with the antenna, ensuring constant orientation towards Earth. A secondary transceiver was connected to both a medium-gain and an omni-directional antenna for backup communication. Each transceiver operated at 8 watts, transmitting data using S-band frequencies: 2110 MHz for uplink and 2292 MHz for downlink. NASA's Deep Space Network tracked the spacecraft's signals, and a convolutional encoder on the spacecraft allowed communication errors to be corrected by receiving stations on Earth.

The data transmission rate at launch was 256 bits per second, which decreased by approximately 1.27 millibits per second each day due to increasing distance from Earth. Much of the mission's computations were conducted on Earth, with commands uploaded to Pioneer 10. The spacecraft could store up to five commands at a time from a set of

222 possible instructions. It also carried a data storage unit capable of recording 6,144 bytes of information collected by its scientific instruments, which was prepared for transmission to Earth by the digital telemetry unit.

Pioneer 10 was equipped with a suite of instruments designed to study the environment of the outer Solar System and Jupiter. These instruments provided critical data on magnetic fields, solar wind, cosmic rays, and the Jovian system.

Helium Vector Magnetometer (HVM): Mounted on a 6.6-meter boom to minimize interference from the spacecraft's magnetic field, this instrument mapped Jupiter's magnetic field and evaluated how the solar wind interacted with the planet. It measured the fine structure of interplanetary magnetic fields.

Principal Investigator: Edward Smith (JPL)

Quadrispherical Plasma Analyzer: Positioned to peer through an opening in the large dish-shaped antenna, this instrument detected particles from the solar wind, originating from the Sun.

Principal Investigator: Aaron Barnes (NASA Ames)

Charged Particle Instrument (CPI): This device detected cosmic rays, particles that permeate the Solar System, providing insight into the composition and energy of these high-energy particles.

Principal Investigator: John Simpson (University of Chicago)

Cosmic Ray Telescope (CRT): The CRT gathered data on cosmic ray particles' composition and energy ranges, adding to scientists' understanding of cosmic phenomena.

Principal Investigator: Frank B. McDonald (NASA Goddard Space Flight Center)

Geiger Tube Telescope (GTT): This instrument surveyed the intensities, energy spectra, and angular distributions of electrons and protons, especially within Jupiter's radiation belts.

Principal Investigator: James A. Van Allen (University of Iowa)

Trapped Radiation Detector (TRD): The TRD was designed to measure electrons and protons in Jupiter's radiation belts. It utilized a Cerenkov counter, an electron scatter detector, and a minimum ionizing detector to record particles in various energy ranges.

Principal Investigator: R. Fillius (University of California, San Diego)

Meteoroid Detectors: Twelve pressurized cell panels mounted on the back of Pioneer 10's main dish antenna recorded impacts from small meteoroids during the spacecraft's journey.

Principal Investigator: William Kinard (NASA Langley Research Center)

Asteroid/Meteoroid Detector (AMD): This instrument, consisting of four non-imaging telescopes, tracked particles ranging from dust to large asteroids, contributing to a broader understanding of the interplanetary environment.

Principal Investigator: Robert Soberman (General Electric Company)

Ultraviolet Photometer: The UV Photometer detected ultraviolet light to measure the concentrations of hydrogen and helium, both in space and on Jupiter.

Principal Investigator: Darrell Judge (University of Southern California)

Imaging Photopolarimeter (IPP): This imaging system relied on the spacecraft's spin to capture narrow strips of images in red and blue light. These strips were later processed to form visual images of Jupiter and its moons.

Principal Investigator: Tom Gehrels (University of Arizona)

Infrared Radiometer: The infrared radiometer provided critical information on the cloud temperatures and heat output from Jupiter, helping scientists better understand the planet's atmospheric dynamics.

Principal Investigator: Andrew Ingersoll (California Institute of Technology)

Together, these instruments allowed Pioneer 10 to complete its primary mission of exploring Jupiter and the asteroid belt while also paving the way for future deep-space exploration. The data collected significantly advanced our understanding of the outer Solar System, including the nature of cosmic rays, the solar wind, and the composition of Jupiter's magnetic and radiation environments.

Pioneer 10, a mission of monumental significance, was launched by NASA on March 3, 1972, at 01:49:00 UTC (March 2, 8:49 p.m. EST) from Space Launch Complex 36A in Florida. The spacecraft was carried aboard an Atlas-Centaur rocket, a powerful launch vehicle that included a specially developed third stage—the Star-37E (TE-M-364-4)—which provided an additional 6,800 kg of thrust. This solid-fuel stage was designed specifically for the Pioneer missions, ensuring that the spacecraft would have the velocity necessary to escape Earth's gravitational pull and begin its journey to the outer reaches of the Solar System. The Star-37E stage also helped spin up Pioneer 10 to an initial rate of 30 rpm.

Twenty minutes after launch, Pioneer 10 deployed its three booms, which slowed its spin to a more manageable 4.8 rpm—a rate it maintained throughout its voyage. The launch vehicle accelerated Pioneer 10 to a blistering speed of 51,682 km/h (32,114 mph), making it the fastest human-made object at that time. After 17 minutes of powered flight, the spacecraft achieved its intended velocity and course.

Within ninety minutes of launch, Pioneer 10 passed beyond Earth's radiation belts and entered interplanetary space. During this time, the spacecraft extended its high-gain antenna and activated several instruments for initial testing. Just eleven hours after launch, Pioneer 10 passed the Moon and, within two days, began switching on its scientific instruments, starting with the cosmic ray telescope. By the tenth day of its mission, all onboard instruments were fully operational, marking the beginning of its ambitious scientific journey.

During the first seven months of its journey, Pioneer 10 underwent three course corrections to fine-tune its trajectory toward Jupiter. During this period, its instruments were calibrated, with the photometers conducting observations of Jupiter and the zodiacal light, while other instruments measured cosmic rays, magnetic fields, and solar wind conditions. Despite an early failure of the Canopus sensor—a key device for spacecraft orientation—the two Sun sensors provided sufficient data to maintain the spacecraft's orientation toward Earth.

While traveling through the vast interplanetary medium, Pioneer 10 achieved several groundbreaking milestones. It became the first spacecraft to detect interplanetary helium atoms, a discovery that contributed to the study of the Solar System's composition. Additionally, Pioneer 10 recorded high-energy ions of aluminum and sodium within the solar wind, further enhancing our understanding of solar phenomena.

In early August 1972, Pioneer 10 captured a significant solar event—registering a solar shock wave caused by a large solar flare. At the time, the spacecraft was at 2.2 astronomical units (AU), or approximately 330 million kilometers (200 million miles), from the Sun. The data collected during this event provided valuable insights into the behavior of solar shock waves and their impact on the heliosphere.

On July 15, 1972, Pioneer 10 became the first spacecraft to enter the asteroid belt—a region of space located between the orbits of Mars and Jupiter. The asteroid belt had long been uncertain for mission planners, as there were concerns about the potential danger posed by collisions with debris. However, the closest approach Pioneer 10 made to any known asteroid was 8.8 million kilometers (5.5 million miles) from 307 Nike on December 2, 1972. This distance ensured that the spacecraft remained safe while continuing its voyage through this unexplored territory.

As Pioneer 10 traversed the asteroid belt, its instruments gathered invaluable data on the nature of the belt itself. Contrary to expectations, the spacecraft found a notable deficiency of particles smaller than one micrometer in diameter compared to Earth's vicinity. In contrast, the density of dust particles ranging from 10 to 100 micrometers remained relatively constant throughout its journey. However, a threefold increase in density was observed for particles between 100 micrometers and 1 millimeter in size as Pioneer 10 passed through the belt.

Despite these findings, the spacecraft did not encounter any fragments larger than one millimeter, confirming that larger particles are exceedingly rare in the asteroid belt. This realization greatly alleviated earlier fears of collision with substantial debris. After navigating the asteroid belt without incident, Pioneer 10 safely emerged on the other side by February 15, 1973, marking another historic milestone in its mission to Jupiter.

Pioneer 10's historic encounter with Jupiter began on November 6, 1973, when the spacecraft was 25 million kilometers (16 million miles) from the gas giant. Initial testing of the imaging system commenced at this distance, with data successfully relayed to NASA's Deep Space Network. Over the next sixty days, Pioneer 10 was programmed with 16,000 commands to control the intricate operations of its flyby. By November 8, it had crossed the orbit of the outer moon Sinope, marking its first passage through Jupiter's extensive satellite system.

On November 16, Pioneer 10 reached the bow shock of Jupiter's magnetosphere, a boundary where the solar wind abruptly slows from 451 km/s (280 mi/s) to 225 km/s (140 mi/s), signifying the spacecraft's entry into Jupiter's magnetic domain. A day later, it passed through the magnetopause, the outer boundary of Jupiter's magnetosphere, and began detecting the planet's magnetic field, which was found to be inverted compared to Earth's. By November 29, Pioneer 10 had crossed the orbits of all of Jupiter's outermost moons and was functioning flawlessly as it approached the planet.

The spacecraft's imaging photopolarimeter captured a series of red and blue images as Pioneer 10's rotation swept the camera's field of view across Jupiter's surface. These images were later combined with synthetic green to create full-color representations of the planet. On November 26, the first twelve images were transmitted back to Earth, revealing details of Jupiter that far exceeded the best ground-based observations. These real-time images were displayed for the public, and Pioneer 10's successful media presentation earned the program an Emmy award. Despite geometric distortions from the spacecraft's motion, computer processing later corrected these issues. By the end of the encounter, Pioneer 10 had sent over 500 images back to Earth.

Pioneer 10's trajectory took it along Jupiter's magnetic equator, where ion radiation was most concentrated. The spacecraft encountered radiation levels that were 10,000 times stronger than Earth's, with electron and proton radiation doses far exceeding expectations. As Pioneer 10 passed within 20 Jupiter radii (20 RJ) of the planet, it received an integrated dose of 200,000 rads from electrons and 56,000 rads from protons—a radiation level so intense that it was ten times what the spacecraft's designers had predicted. These doses were far beyond what a human could survive, underscoring the extreme conditions of the Jovian environment.

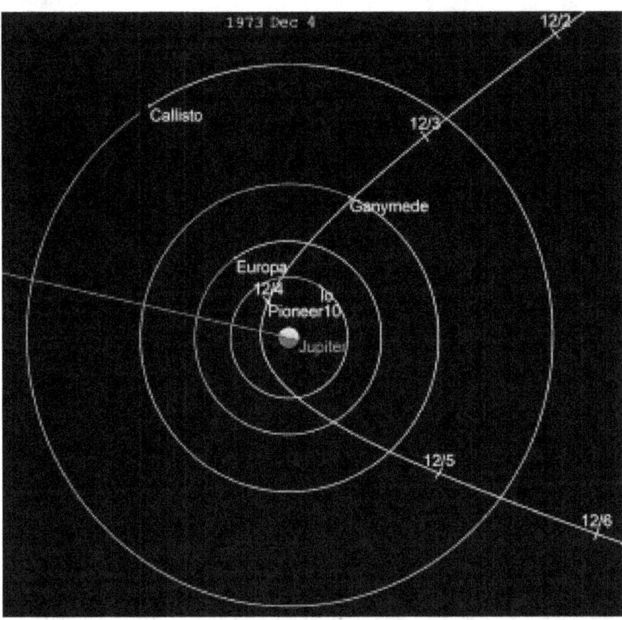

On December 3, as the spacecraft approached its closest point to Jupiter, the high levels of radiation began to interfere with its systems, generating false commands. Mission controllers implemented contingency plans to correct most of these issues, but some data, including images of Io and close-ups of Jupiter, were lost. Despite these challenges, Pioneer 10 successfully transmitted detailed images of Jupiter's moons Ganymede and Europa. The image of Ganymede revealed low-albedo features near its south pole, while Europa, though farther away, displayed some distinct surface features.

Pioneer 10's trajectory was designed to pass behind Io, Jupiter's volcanic moon, allowing the spacecraft to measure the refractive effects of Io's ionosphere on radio transmissions. These measurements revealed that Io's ionosphere extended 700 kilometers (430 miles) above its surface, with electron densities ranging from 60,000 electrons per cubic centimeter on the day side to 9,000 on the night side. Additionally, the spacecraft made a surprising discovery: Io was surrounded by a massive cloud of hydrogen, extending 805,000 kilometers (500,000 miles) from the moon. A smaller hydrogen cloud, approximately 110,000 kilometers (68,000 miles) in diameter, was also detected near Europa.

These unexpected findings regarding the moons of Jupiter offered crucial insights into the complex environments surrounding the gas giant and laid the groundwork for future missions to explore these enigmatic worlds.

As Pioneer 10 approached its closest point to Jupiter on December 3, 1973, it reached a velocity of 132,000 km/h (82,000 mph) and passed within 132,252 kilometers (82,178 miles) of the planet's outer atmosphere. During this flyby, the spacecraft captured close-up images of Jupiter's Great Red Spot and the planet's terminator, providing humanity with its first detailed look at these iconic features. Communication with the spacecraft was briefly lost as it passed behind the planet, but when contact resumed, Pioneer 10 continued transmitting vital data.

One of the most significant findings was derived from radio occultation measurements, which provided a detailed profile of Jupiter's atmospheric structure. These data revealed a temperature inversion between atmospheric pressures of 10 and 100 millibars, with temperatures ranging from −133 to −113 °C (−207 to −171 °F) at the 10-mbar level and between −183 and −163 °C (−297 to −261 °F) at 100 millibars. Pioneer 10's infrared observations confirmed that Jupiter radiated more heat than it received from the Sun, a phenomenon long suspected by scientists.

After completing its closest approach, Pioneer 10 transmitted crescent images of Jupiter as it moved away. As it exited Jupiter's magnetosphere, the spacecraft encountered the bow shock 17 times due to the dynamic interaction between Jupiter's magnetic field and the solar wind. Pioneer 10's successful traversal of this region marked the completion of its mission objectives at Jupiter, but the spacecraft's journey was far from over.

The encounter with Jupiter was more than just a scientific triumph—it also allowed NASA to use the planet's immense gravitational field for a slingshot maneuver, propelling Pioneer 10 out of the Solar System. This maneuver was the first of its kind and became a model for future deep-space missions. Though it was not part of the original mission proposal, this extended mission was planned before launch, with the goal of exploring the far reaches of the Solar System and eventually interstellar space.

As Pioneer 10 continued its journey beyond Jupiter, it became the first spacecraft to enter the outer Solar System and paved the way for humanity's exploration of the distant planets and beyond.

Pioneer 10 continued its historic journey after its encounter with Jupiter, entering the far reaches of the Solar System. In 1976, it crossed the orbit of Saturn, followed by the orbit of Uranus in 1979. On June 13, 1983, Pioneer 10 became the first human-made object to leave the vicinity of the major planets by crossing the orbit of Neptune. This marked a significant milestone, as the spacecraft ventured beyond the outer boundaries of the planetary system into the realm of deep space.

The spacecraft's mission officially ended on March 31, 1997, when it had reached 67 astronomical units (AU), approximately 10 billion kilometers (6.2 billion miles) from the Sun. However, Pioneer 10 continued to transmit coherent data after this date, and its weak signal was tracked by the Deep Space Network (DSN). The tracking helped train flight controllers in the challenging process of acquiring deep-space radio signals. Additionally, an Advanced Concepts study applied chaos theory to extract coherent data from the fading signal.

The final successful telemetry transmission from Pioneer 10 was received on April 27, 2002, at 80.22 AU from Earth. Subsequent signals were too weak to provide usable data. On January 23, 2003, a very faint signal was detected from the spacecraft, which by then had reached 12 billion kilometers (80 AU; 7.5 billion miles) from Earth. After this, NASA made several attempts to contact the spacecraft, including a final attempt on March 4, 2006, when the spacecraft's antenna was last correctly aligned with Earth. No response was received, leading NASA to conclude that the spacecraft's radioisotope thermoelectric generators (RTGs) had likely decayed below the threshold needed to power the transmitter.

Timeline of Milestones

March 3, 1972: Pioneer 10 launched.

July 15, 1972: Entered the asteroid belt.

December 3-4, 1973: Encountered the Jovian system and completed its closest approach to Jupiter.

February 10, 1975: The U.S. Post Office issued a commemorative stamp featuring Pioneer 10.

April 25, 1983: Crossed the orbit of Pluto, which at the time was still considered a planet.

June 13, 1983: Crossed Neptune's orbit and became the first spacecraft to leave the vicinity of the Solar System's major planets.

March 31, 1997: The mission officially ended, but telemetry continued.

February 17, 1998: Voyager 1 overtook Pioneer 10 as the most distant human-made object from the Sun.

April 27, 2002: The last successful telemetry data was received from Pioneer 10.

January 23, 2003: The final signal was received from the spacecraft.

March 4, 2006: The last attempt to contact Pioneer 10, without success.

Even after contact was lost, Pioneer 10 continued on its trajectory into interstellar space. As of June 2024, Pioneer 10 is approximately 137.3 AU (20.5 billion kilometers; 12.8 billion miles) from Earth, and 136.3 AU (20.4 billion kilometers; 12.7 billion miles) from the Sun. Sunlight takes approximately 18.9 hours to reach the spacecraft, with the Sun appearing as a faint magnitude −16.0 object from its distant location. Pioneer 10 is currently traveling in the direction of the constellation Taurus.

In July 2023, Voyager 2 overtook Pioneer 10, making it the third farthest spacecraft from the Sun after Voyager 1 and Voyager 2. Pioneer 10 will continue to drift through interstellar space, joining its sister craft, Pioneer 11, and the two Voyager spacecraft. Its trajectory is expected to take it in the general direction of the star Aldebaran, located about 68 light-years away. However, even at its current velocity, it would take over two million years for Pioneer 10 to reach this distant star, assuming no relative motion between the spacecraft and Aldebaran.

In approximately 90,000 years, before reaching Aldebaran, Pioneer 10 will pass within 0.75 light-years of the K-type star HIP 117795. This will be the closest stellar encounter of any of the Pioneer, Voyager, or New Horizons spacecraft as they journey beyond the Solar System. These distant voyages are a testament to humanity's first steps into deep space, and Pioneer 10's legacy will endure as it travels through the cosmos, a silent ambassador of Earth's early exploration efforts.

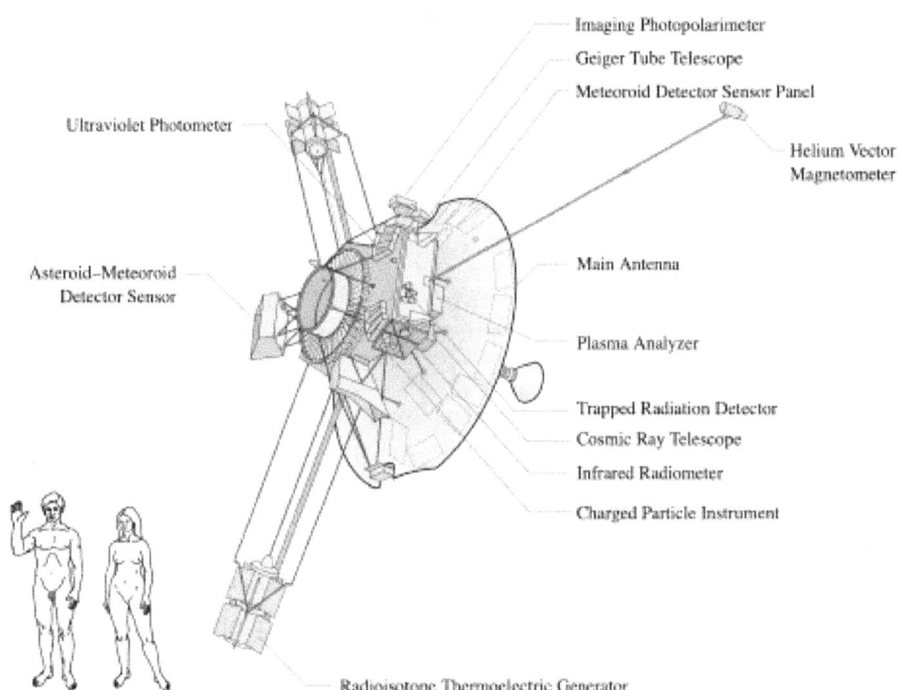

Chapter 7 - The Era of Cooperation: Apollo-Soyuz and Space Stations

Soyuz 19

Apollo–Soyuz Test Project

First multinational human-crewed mission

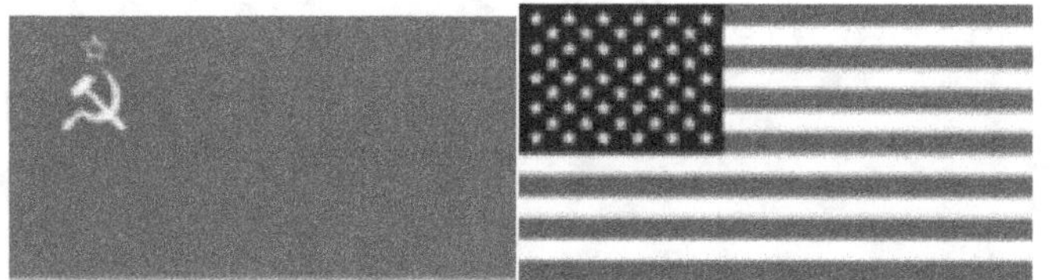

The Apollo-Soyuz Test Project (ASTP) was a historic milestone in space exploration, marking the first joint crewed mission between the United States and the Soviet Union. Conducted in July 1975, this mission symbolized a significant step toward détente during the height of the Cold War, a period marked by intense rivalry between the two superpowers. Millions of people around the world watched as an American Apollo spacecraft docked with a Soviet Soyuz capsule, a moment that captured the imagination of the world and embodied the possibility of peaceful cooperation between adversaries.

The ASTP mission, known as Eksperimentalniy Polyot Soyuz-Apollo (EPAS) in the Soviet Union, was officially designated Soyuz 19 by the Soviets. For the Americans, this mission represented the final flight of an Apollo spacecraft. The unnumbered Apollo vehicle was a leftover module from the canceled Apollo lunar missions and marked the end of an era for the United States' pioneering lunar program.

The American crew for the Apollo-Soyuz mission consisted of three astronauts:

Commander Thomas P. Stafford: A veteran of three previous spaceflights, including Gemini 6A, Gemini 9A, and Apollo 10, Stafford commanded the Apollo spacecraft in what would be his fourth and final spaceflight. His experience made him a natural choice for leading this historic mission.

Command Module Pilot Vance D. Brand: Making his first spaceflight, Brand was responsible for piloting the Apollo Command Module. His calm demeanor and technical expertise played a crucial role in ensuring the smooth operation of the mission.

Astronaut Thomas P. Stafford (standing on left), commander of the American crew; cosmonaut Aleksey A. Leonov (standing on right), commander of the Soviet crew; astronaut Donald K. Slayton (seated on left), docking module pilot of the American crew; astronaut Vance D. Brand (seated center), command module pilot of the American crew; and cosmonaut Valeriy N. Kubasov (seated on right), engineer on the Soviet crew. The crew members wear the same mission patch, but oriented to reflect "Soyuz-Apollo" or "Apollo-Soyuz", as the program was called in their respective countries.

Docking Module Pilot Donald K. "Deke" Slayton: One of the original Mercury Seven astronauts selected in 1959, Slayton's participation in this mission marked his only spaceflight. Grounded for over a decade due to a medical condition, Slayton's persistence and contributions to NASA as head of the Astronaut Office finally paid off when he was cleared to fly in 1972.

The Apollo crew also had a backup team:

Commander Alan Bean, who had flown on Apollo 12 and Skylab.

Command Module Pilot Ronald Evans, a veteran of Apollo 17.

Docking Module Pilot Jack Lousma, who had flown on Skylab 3.

Notably, Jack Swigert, the command module pilot for Apollo 13, had originally been assigned to the ASTP prime crew. However, he was removed from the mission due to his involvement in the Apollo 15 postal covers incident.

The Soviet crew for the Soyuz spacecraft consisted of:

Commander Alexei Leonov: Leonov, one of the Soviet Union's most famous cosmonauts, was known for being the first person to walk in space during the Voskhod 2 mission in March 1965. The Apollo-Soyuz mission would be his second and final spaceflight.

Flight Engineer Valery Kubasov: Kubasov, a veteran of the Soyuz 6 mission, served as the flight engineer for the Soviet spacecraft. Like Leonov, this mission marked his second spaceflight.

The Soviet backup crew included Commander Anatoly Filipchenko, a veteran cosmonaut from Soyuz 7 and Soyuz 16 and Flight Engineer Nikolai Rukavishnikov, who had flown on Soyuz 10.

The Apollo-Soyuz mission was notable for several personal milestones among its participants. For Deke Slayton, the mission represented his long-awaited first and only spaceflight after years of being grounded due to medical concerns. Slayton had been one of NASA's original Mercury Seven astronauts but was removed from flight status in the early 1960s due to a heart condition. After regaining his flight status in 1972, the ASTP mission allowed him to finally achieve his dream of flying in space.

Another notable figure associated with the mission was Jack Swigert, who had originally been assigned as the command module pilot for the ASTP crew. Swigert, famous for his role in the Apollo 13 mission, was removed from the crew before the official announcement due to his involvement in the Apollo 15 postal covers incident, a controversy surrounding unauthorized postal stamps flown aboard that mission.

For the Soviet side, Alexei Leonov's participation in the Apollo-Soyuz mission was a fitting capstone to a storied career. Leonov had made history with his spacewalk in 1965, and this mission would mark the final chapter of his remarkable contributions to space exploration.

The Apollo-Soyuz mission brought together these astronauts and cosmonauts in a historic partnership that symbolized not only scientific cooperation but also a significant political act during the Cold War.

The Apollo-Soyuz mission was far more than a symbolic gesture of peace. It provided crucial technical experience for future collaborative space efforts, laying the groundwork for later joint missions such as the Shuttle-Mir program and the International Space Station (ISS). The coordination between the two nations' space programs in areas like spacecraft design, life-support systems, and scientific research was invaluable for future international space missions.

For the United States, ASTP marked the end of an era and a pause in human spaceflight. After this mission, the U.S. would not send another crewed spacecraft into orbit for nearly six years, until the Space Shuttle program commenced with the launch of Columbia on April 12, 1981. It was also the last time a U.S. spacecraft with a capsule design would fly until Crew Dragon Demo-2 in May 2020.

The Apollo-Soyuz Test Project (ASTP) was a landmark mission born from the broader political movement known as détente, which aimed to ease tensions between the two Cold War superpowers, the United States and the Soviet Union. By the early 1970s, U.S. President Richard Nixon and Soviet Premier Alexei Kosygin had initiated steps toward improving bilateral relations, recognizing the potential for cooperation amidst the enduring rivalry. On May 24, 1972, the two leaders signed an agreement in Moscow that laid the foundation for a cooperative space mission. This was a significant diplomatic breakthrough, occurring at a time when the United States was still embroiled in the Vietnam War, and relations between the two nations were strained.

The relationship between the United States and the Soviet Union during the space race had been characterized by intense competition, with each side striving for supremacy in space exploration. The Soviet press, for instance, was critical of U.S. space efforts, including the Apollo program. During the launch of Apollo 14 in 1971, Soviet newspapers ran scathing headlines, such as, "the armed intrusion of the United States and Saigon puppets into Laos is a shameless trampling underfoot of international law." Despite Soviet leader Nikita Khrushchev's earlier push for peaceful coexistence in 1956, genuine collaboration in space exploration would not materialize until much later.

The first steps toward such cooperation began in the early 1960s. After astronaut John Glenn's orbital flight in 1962, an exchange of letters between U.S. President John F. Kennedy and Soviet Premier Khrushchev initiated discussions on potential joint space missions. NASA Deputy Administrator Hugh Dryden and Soviet scientist Anatoly Blagonravov led the talks, which culminated in the Dryden-Blagonravov agreement in October 1962, during the height of the Cuban Missile Crisis. This agreement called for cooperation in areas such as satellite tracking, studies of Earth's magnetic field, and the exchange of weather data. Despite this initial progress, the assassination of Kennedy in 1963 and Khrushchev's ousting in 1964 stalled further collaboration, and the competition between the nations' space programs intensified.

By 1971, the rivalry in space had reached new heights. The Soviet Union launched the world's first space station, Salyut 1, on April 19, 1971, while the United States continued its Apollo program, achieving its third Moon landing with Apollo 14 earlier that year. Both sides remained reluctant to acknowledge the other's achievements, and distrust between the two nations persisted. The United States criticized the Soviet approach to spacecraft design, particularly the automation of systems in the Soyuz capsule, which was built to be controlled largely from the ground. American engineers, including Christopher C. Kraft, Director of the Johnson Space Center, expressed concerns about the lack of redundant systems in the Soyuz, contrasting it with the Apollo spacecraft, which relied heavily on human operation and had multiple fail-safes. NASA's astronauts were also skeptical of the Soyuz design, preferring the hands-on control offered by their own spacecraft.

Despite these technical differences and political tensions, a willingness to collaborate began to emerge. Between June and December 1971, engineers from both nations met in Houston and Moscow to discuss the feasibility of a joint mission. These discussions focused on designing a docking mechanism that would allow an American Apollo spacecraft to connect with a Soviet Soyuz in orbit. The result was the Androgynous Peripheral Attach System (APAS), a docking system that could allow either spacecraft to play an active or passive role during the docking process. This critical development paved the way for the Apollo-Soyuz mission.

As the Vietnam War drew to a close and relations between the superpowers improved, the prospects for a cooperative space mission became more favorable. By April 1972, both nations had signed the Agreement Concerning Cooperation in the Exploration and Use of Outer Space for Peaceful Purposes, committing to the Apollo-Soyuz Test Project, set to launch in 1975. This mission would become a powerful symbol of détente, representing a thaw in the longstanding rivalry between the United States and the Soviet Union.

Soviet leader Leonid Brezhnev highlighted the mission's significance, stating, "The Soviet and American spacemen will go up into outer space for the first major joint scientific experiment in the history of mankind. They know that from outer space, our planet looks even more beautiful. It is big enough for us to live peacefully on it, but it is too small to be threatened by nuclear war." This sentiment underscored the mission's political importance, which sought to solidify the improving relations between the two nations.

The Apollo-Soyuz Test Project was particularly significant for the Soviet Union, as it marked the first time the USSR had opened the details of its space program to the world. The launch of Soyuz 19, the Soviet spacecraft used in the mission, was televised live—a major departure from the secrecy that had typically surrounded Soviet space missions. Additionally, for the first time, a foreign crew, the American astronauts, were granted access to a Soviet spacecraft and its training facilities, signaling an unprecedented level of transparency in the Soviet space program.

The Apollo-Soyuz Test Project (ASTP) was not universally welcomed, as it sparked mixed reactions in both the United States and the Soviet Union. In the United States, some critics feared that the mission granted the Soviet Union an unwarranted level of recognition, elevating their space program to the same status as NASA's. For many Americans, NASA's successes with the Apollo Moon landings represented a pinnacle of space exploration, and they saw the joint mission as an unnecessary concession that could obscure the significant technological gap between the two space agencies. Skeptics also worried that the appearance of peaceful cooperation might lead the public to overlook the deeper, unresolved political tensions that still simmered between the two superpowers.

In the Soviet Union, similar doubts emerged. Some Soviet publicists were wary of the partnership with the U.S., and they dismissed American critics of the mission as "demagogues who stand against scientific cooperation with the USSR." This reflected the ideological suspicion that still lingered within both governments, despite the diplomatic progress made through détente. Working closely with an ideological rival during the Cold War was fraught with political complexities, and some in both nations questioned whether the benefits of scientific collaboration outweighed the risks of ideological compromise.

The Apollo-Soyuz mission was not just a political challenge but also a technical one. The American and Soviet spacecraft reflected two distinct design philosophies, which led to initial skepticism between the two teams. The Apollo spacecraft, developed by NASA, was built for manual operation, requiring highly trained astronauts to control critical systems during flight. In contrast, the Soviet Soyuz capsule was designed to be highly automated, relying more on ground control for navigation and operations. This difference in approach reflected the Soviet focus on minimizing the potential for human error, while the U.S. favored astronaut autonomy and decision-making in space.

Overcoming these differences required intense collaboration between engineers from both nations. Joint meetings were held to resolve technical issues, particularly the design of the docking mechanism that would allow the two spacecraft to connect in orbit. The resulting Androgynous Peripheral Attach System (APAS) was a triumph of engineering that enabled either spacecraft to be the active or passive participant in the docking process, ensuring flexibility and safety.

Beyond the technical challenges, ASTP broke new ground in terms of transparency. The mission marked the first time a Soviet space launch was broadcast live to an international audience. Viewers around the world were able to follow not only the launch and docking of the two spacecraft but also the interactions between the Soviet and American crews in space. This level of openness was unprecedented for the Soviet space program, which had traditionally been shrouded in secrecy, especially regarding its failures. Allowing international viewers to witness the mission was a significant step toward fostering a sense of cooperation and mutual understanding between the two nations.

Despite the initial concerns and political challenges, the Apollo-Soyuz mission set a lasting precedent for international cooperation in space. The technical and diplomatic groundwork laid by ASTP proved invaluable in subsequent collaborations, such as the Shuttle-Mir program in the 1990s and the ongoing work aboard the International Space Station (ISS). The mission demonstrated that even during times of geopolitical tension, space exploration could serve as a powerful platform for cooperation, advancing scientific knowledge while fostering international goodwill.

ASTP became a symbol of the possibilities for peaceful collaboration in space, a reminder that the cosmos' vastness could transcend Earth's divisions. By working together to solve complex technical problems and conducting joint scientific experiments in space, the United States and the Soviet Union showed that cooperation was not only possible but also beneficial. The success of ASTP paved the way for the future of international space exploration, where the pursuit of knowledge and discovery could bridge even the most profound political divides.

U.S. President Richard Nixon and Soviet Premier Alexei Kosygin (seated) sign an agreement in Moscow paving the way for the Apollo–Soyuz mission, May 1972.

The Apollo-Soyuz Test Project (ASTP) crew represented a significant collaboration between the United States and the Soviet Union, with both sides contributing highly experienced astronauts and cosmonauts. The American crew consisted of three astronauts, while the Soviet crew included two cosmonauts. Together, these individuals carried out the first joint international space mission, symbolizing the thawing of Cold War tensions.

The Apollo-Soyuz Test Project (ASTP) marked a significant milestone in space exploration, symbolizing the first international human spaceflight mission. It involved the docking of an American Apollo Command and Service Module (CSM) with a Soviet Soyuz 7K-TM spacecraft, demonstrating the feasibility of international cooperation in space despite Cold War tensions. The mission took place in July 1975 and was hailed as a step toward peaceful collaboration between the two superpowers.

The Soyuz 7K-TM spacecraft, designated Soyuz 19 for this mission, carried the simple call sign "Soyuz" throughout the joint operation. On the American side, the Apollo CSM similarly bore the call sign "Apollo." Although some media and NASA unofficially referred to this mission as "Apollo 18," this should not be confused with the canceled lunar mission of the same name.

The Apollo spacecraft was launched using a Saturn IB rocket and carried a specially designed docking module, used solely for this mission. The module served as both an airlock and an adapter, addressing the difference in the atmospheric pressures of the two spacecraft. The Apollo used a pure oxygen environment pressurized at 5 psi (34 kPa), while the Soyuz maintained a nitrogen-oxygen mix at standard sea-level pressure (15 psi or 100 kPa). The docking module enabled the Apollo crew to safely dock with the Soyuz, which was equipped with an Androgynous Peripheral Attach System (APAS) collar, jointly developed by NASA and Soviet engineers. The Apollo docking module was retrieved from the upper stage of the Saturn IB rocket after launch, in a procedure similar to that used during the Lunar Module retrieval in earlier Apollo missions.

Commanding the American crew were seasoned astronauts Tom Stafford, Vance Brand, and Deke Slayton. Stafford, a brigadier general in the U.S. Air Force, was flying his fourth mission, having previously come within eight nautical miles of the lunar surface as commander of Apollo 10. Slayton, one of the original Mercury Seven astronauts, had waited 16 years for his first spaceflight after being grounded due to an irregular heartbeat. His participation in ASTP came after a lengthy medical rehabilitation program, making him the oldest person to fly in space at the time. Brand, the least experienced astronaut of the three, had been a backup command module pilot for Apollo 15 and had trained extensively for a potential Skylab rescue mission.

The Soviet crew consisted of Alexei Leonov and Valery Kubasov. Leonov, already famous for being the first human to conduct a spacewalk during the Voskhod 2 mission in 1965, commanded the Soyuz 19 spacecraft. Kubasov, an experienced cosmonaut, had previously flown on Soyuz 6 in 1969, where he conducted early space manufacturing experiments. Both Leonov and Kubasov had been slated to fly aboard the ill-fated Soyuz 11 mission in 1971 but were grounded due to health concerns, narrowly avoiding the fatal depressurization incident that claimed the lives of the Soyuz 11 crew.

The Soyuz 7K-TM used for ASTP was a modified version of the post-Soyuz 11 two-man spacecraft. Unlike earlier models, it was equipped with solar panels and operated at a reduced cabin pressure of 10.2 psi (70 kPa), which eased transfers between the Soviet and American crews. In total, six ASTP-class Soyuz spacecraft were built, including the one used for the mission. Prior to ASTP, two uncrewed spacecraft were launched as Kosmos satellites, and a crewed Soyuz 16 flight served as a dress rehearsal for the docking procedure. A backup Soyuz was kept ready on the launch pad but was not used during the mission.

On July 15, 1975, Soyuz 19 and Apollo launched within seven and a half hours of each other. Two days later, on July 17, the spacecraft docked in orbit, and American astronaut Tom Stafford and Soviet cosmonaut Alexei Leonov exchanged the first international handshake in space. This historic gesture was initially expected to occur over Bognor Regis, England, but delays resulted in it taking place over Metz, France. The joint crews received messages of congratulations from U.S. President Gerald Ford and Soviet General Secretary Leonid Brezhnev.

While the spacecraft were docked, the two crews engaged in a variety of joint scientific experiments and cultural exchanges. Among the scientific studies was a notable experiment led by American embryologist Jane Oppenheimer, who investigated the effects of weightlessness on fish eggs at various stages of development. In addition to the scientific activities, the astronauts exchanged flags, gifts, and even music. The Soviet crew introduced the Americans to "Tenderness" by Maya Kristalinskaya, while the Americans played "Why Can't We Be Friends?" by the band War. There were also joint meals and visits to each other's spacecraft, with conversations in both Russian and English.

One of the highlights of the mission was a docking and redocking maneuver, during which the two spacecraft reversed roles, with Soyuz becoming the "active" ship. After 44 hours of joint operations, the two spacecraft separated, but not before performing an artificial solar eclipse using the Apollo spacecraft to allow the Soyuz crew to photograph the solar corona. The Soviet crew remained in space for two more days, while the American crew conducted Earth observation experiments for an additional five days.

Mission control center in Houston during ASTP

The Apollo-Soyuz Test Project (ASTP) was hailed as a significant success, both technically and as a symbol of Cold War détente. However, the mission was not without challenges. During reentry and splashdown of the Apollo spacecraft, the crew encountered a serious issue when they were accidentally exposed to toxic fumes. Unignited hypergolic propellants from the Reaction Control System (RCS) leaked and were vented from the spacecraft, entering the cabin through an air intake. This occurred because the RCS had been inadvertently left on during descent. As the Apollo drew in outside air, toxic fumes, including monomethylhydrazine and nitrogen tetroxide, were sucked into the cabin, causing immediate distress for the crew.

Vance Brand, the command module pilot, briefly lost consciousness due to the fumes. Fortunately, mission commander Tom Stafford acted quickly, retrieving emergency oxygen masks. He assisted Brand and fellow astronaut Deke Slayton, helping them don the masks and avoid further exposure. The crew was subsequently hospitalized in Honolulu, Hawaii, for two weeks as they recovered from the incident. Brand later took responsibility for the mishap, attributing the error to the high noise levels in the cabin during reentry, which prevented him from hearing Stafford's call to switch off the RCS. The oversight delayed the automatic shutdown of the system and the deployment of the drogue parachutes, allowing the toxic propellants to leak into the spacecraft.

This mission marked the final flight of an Apollo spacecraft. As soon as the Apollo launched, preparations were underway to convert the launch facilities at Kennedy Space Center for the upcoming Space Shuttle program. Launch Complex 39B and the Vehicle Assembly Building were modified for the shuttle, while Launch Complex 39A, used for previous Apollo missions, had already been closed following the launch of Skylab.

The Apollo-Soyuz mission also left a lasting technical legacy. The docking collar used in the mission, known as the Androgynous Peripheral Attach System (APAS), was a key development in international space docking technology. Although the APAS design used for Apollo-Soyuz was not mechanically compatible with later systems, it provided the foundation for future docking systems. A modified version, the APAS-89, was launched as part of the Soviet Mir space station's Kristall module. Initially intended for the Soviet Buran space shuttle, the APAS-89 was ultimately used for the next Russian-American docking mission in 1995, when Space Shuttle Atlantis docked with Mir during STS-71.

The APAS system continued to evolve, playing a critical role in future space cooperation. The American Space Shuttle program used the APAS-89 docking hardware to connect with both the Mir space station and later the International Space Station (ISS). The Pressurized Mating Adapters (PMAs), equipped with the newer APAS-95 adapters, allowed the ISS to accommodate Space Shuttle missions. While the APAS-95 adapters were compatible with APAS-89, they were no longer androgynous, meaning they could not function as the "active" partner in docking. Today, the first PMA, PMA-1, remains in use on the ISS, connecting the Russian-built Zarya module to the U.S. segment of the station. Thus, the legacy of the APAS system, initiated by Apollo-Soyuz, continues in space operations in 2024.

Apollo-Soyuz was a landmark in international diplomacy, representing the first joint U.S.-Soviet space mission. It was widely anticipated that the mission would either lead to increased international cooperation in space or intensify the competitive nature of space exploration. In reality, it did both. The mission became symbolic of scientific collaboration between two superpowers, yet the political narratives from each nation downplayed the technological capabilities of the other. Soviet media portrayed the mission as evidence of their leadership in space exploration, linking their achievements to Marxist-Leninist ideology. Meanwhile, American reports suggested that the Soyuz spacecraft was technologically inferior to the Apollo system. Despite these competitive narratives, the mission laid the groundwork for future collaboration, leading to programs like Shuttle-Mir and eventually the International Space Station.

However, high-profile cooperation between the United States and the Soviet Union in space diminished after the Apollo-Soyuz mission, caught up in the complex political landscape of the time. Nonetheless, the precedent set by this mission for joint efforts in space exploration persisted, evolving into future collaborations that transcended the Cold War rivalry.

One of the enduring legacies of the Apollo-Soyuz mission was the personal friendship between the two mission commanders, Tom Stafford and Alexei Leonov. Their camaraderie extended beyond the mission, with Leonov becoming the godfather to Stafford's younger children. When Leonov passed away in October 2019, Stafford delivered a heartfelt eulogy at his funeral, a testament to the deep bond they had forged during the mission.

The mission also left its mark on popular culture. In 1977, Soviet astronomer Nikolai Chernykh discovered an asteroid, which he named 2228 Soyuz-Apollo in honor of the historic mission. Additionally, renowned bartender Joe Gilmore from The Savoy Hotel's American Bar created a commemorative cocktail called the 'Link-Up' for the astronauts. Upon hearing of the creation, the astronauts reportedly joked that they wanted the cocktail delivered to space.

The Apollo spacecraft carried the Solar Astronomy Group (SAG) telescope, designed to observe the extreme ultraviolet spectrum. During the mission, the telescope made several important discoveries, identifying two ultraviolet sources: HZ 43 and FEIGE 24, both of which were white dwarf stars. Other celestial objects observed included Proxima Centauri, a red dwarf star, SS Cygni, a binary star system, and Sirius, another binary star. There was also a potential discovery of an unknown object in the Pavo constellation, though later analysis ruled out the suggested candidate star, HD 192273, as being too distant and of an inappropriate spectral class to match the observed object.

The scientific achievements of the Apollo-Soyuz Test Project contributed to a deeper understanding of space phenomena, while its technical and political legacy paved the way for future international cooperation in space exploration. The mission remains a symbol of what can be accomplished when nations set aside differences to work together in the pursuit of scientific progress.

Venera 9

The first spacecraft to orbit Venus (the orbiter)
First view and clear photograph from and of the surface of another planet (the lander)

Venera 9 (Russian: Венера-9), officially designated as 4V-1 No. 660, marked a significant milestone in space exploration as the first spacecraft to orbit Venus and return images from its surface. Launched by the Soviet Union on June 8, 1975, at 02:38 UTC, Venera 9 weighed approximately 4,936 kilograms (10,882 lbs) and consisted of both an orbiter and a lander. This dual-mission aimed to expand humanity's understanding of Venus, a planet often shrouded in mystery due to its dense atmosphere and extreme surface conditions.

The orbiter entered Venusian orbit on October 20, 1975, becoming the first human-made object to do so. Its primary objective was to serve as a communications relay for the lander and conduct scientific observations of Venus's cloud layers and atmospheric composition. Between October 26 and December 25, 1975, the orbiter carried out 17 survey missions, collecting valuable data on Venus's atmosphere and surface.

The orbiter's design featured a cylindrical body flanked by two solar panel wings and equipped with a high-gain parabolic antenna for communications. A bell-shaped propulsion system was mounted at the base of the orbiter, while the lander was housed in a 2.4-meter spherical compartment at the top. This modular design was essential for carrying out both the orbiter's and lander's specific roles.

The orbiter was equipped with a comprehensive suite of scientific instruments designed to study Venus's atmosphere and magnetic field. These included infrared and ultraviolet spectrometers, photo-polarimeters, and a magnetometer. The orbiter also carried radio occultation equipment and imaging devices such as a UV camera and an array of electrostatic analyzers to detect various particles in Venus's upper atmosphere. Together, these instruments painted a detailed picture of Venus's environmental conditions, laying the groundwork for future missions.

On October 22, 1975, the Venera 9 lander became the first spacecraft to transmit images from another planet's surface successfully. The journey to the surface of Venus was a complex and dangerous process, given the planet's extreme atmospheric conditions. The lander, initially encased in a spherical shell, endured the intense heat of atmospheric entry as it decelerated from a speed of 10.7 kilometers per second (6.6 miles per second) to 150 meters per second (490 feet per second). Explosive bolts jettisoned the protective shell, deploying a three-domed parachute at an altitude of 63 kilometers (39 miles) to further reduce speed.

As the lander descended through the thick cloud layers, it relayed data about the Venusian atmosphere back to the orbiter. To mitigate the extreme heat near the surface, the parachute was jettisoned at 50 kilometers (31 miles), and a ring-shaped aerodynamic shield provided the final braking. Upon landing, the base of the lander—a hollow ring—was designed to compress, absorbing the impact of the touchdown on Venus's rocky terrain.

The lander touched down on a steep slope near Beta Regio at coordinates 31.01°N, 291.64°E, an area suspected to be part of a tectonic rift valley known as Aikhylu Chasma. Despite the challenging terrain, the lander began transmitting scientific data almost immediately, revealing a landscape covered in boulders with a smooth, rock-strewn surface. The lander recorded a light level of 14,000 lux, comparable to full daylight on Earth, but with no direct sunlight due to the dense cloud cover.

The Venera 9 lander operated for 53 minutes on the Venusian surface before losing radio contact as the orbiter moved out of range. During this brief but historic mission, the lander measured a surface temperature of 485°C (905°F) and a crushing atmospheric pressure of 9,100 kilopascals (90 atmospheres)—conditions far more extreme than anticipated. The lander also provided the first direct measurements of Venus's atmospheric composition, detecting chemicals such as hydrochloric acid, hydrofluoric acid, bromine, and iodine.

Perhaps Venera 9's most iconic achievement was transmitting black-and-white television images from the surface. These images, although limited to a 180-degree view due to a malfunction with one of the camera lens covers, showed a barren, rock-strewn landscape with no apparent dust or shadows. Rocks measured between 30 and 40 centimeters in size, and they appeared unweathered, suggesting that erosion processes on Venus were either minimal or entirely different from those on Earth.

The lander carried a range of scientific instruments, including temperature and pressure sensors, an accelerometer, and photometers. It also featured mass spectrometers for analyzing the composition of the atmosphere and surface materials, as well as gamma-ray spectrometers and densitometers to study the planet's crust. These instruments provided crucial insights into Venus's geological and atmospheric properties, furthering our understanding of one of Earth's closest neighbors.

The success of Venera 9 paved the way for future Soviet missions to Venus, including Venera 10, which landed shortly after and confirmed many of Venera 9's findings. While the Venera program faced numerous challenges due to the harsh conditions on Venus, these missions contributed significantly to planetary science, offering a rare glimpse of one of the most inhospitable environments in our solar system.

Pioneer 11

First Saturn flyby

Pioneer 11, also known as Pioneer G, was a groundbreaking NASA robotic space probe launched on April 5, 1973. Its mission was to explore the asteroid belt, study the environments of Jupiter and Saturn, and investigate solar wind and cosmic rays. Pioneer 11 became the first probe to encounter Saturn, the second to fly through the asteroid belt, and the second to fly by Jupiter. Eventually, it became one of only five artificial objects to reach an escape velocity sufficient to leave the Solar System. Contact with Pioneer 11 ceased on September 30, 1995, with the last valuable engineering data transmitted on November 24, 1995, marking the end of its illustrious journey through the cosmos.

The Pioneer 11 mission was born from NASA's desire to explore the outer Solar System. In February 1969, NASA approved the twin missions of Pioneer 10 and Pioneer 11, designed explicitly for outer planetary exploration. Pioneer 10, launched in March 1972, served as a precursor, paving the way for Pioneer 11's even more ambitious objectives. Early mission plans targeted exploration of the interplanetary medium beyond Mars and a study of the asteroid belt, both to gain scientific insight and to assess potential hazards for future missions to the outer planets.

However, as planning progressed, Pioneer 11's scope expanded to include a historic encounter with Saturn. The mission objectives now aimed to map Saturn's magnetic field, determine the intensity and structure of its environment, and investigate the interactions between Saturn's system and the solar wind. The probe would also take crucial measurements of Saturn's atmosphere and its largest moon, Titan, as well as conduct radio occultation experiments to probe Saturn's rings and atmospheric structure. These efforts would set the stage for later missions, such as the Mariner Jupiter/Saturn mission, and inform the planning of the famous Voyager program.

Pioneer 11 was constructed by TRW and managed by NASA's Ames Research Center as part of the Pioneer program. The spacecraft, which weighed 259 kilograms, was designed with a hexagonal structure, measuring 36 centimeters deep and equipped with six panels, each 76 centimeters long. It housed the spacecraft's propellant for orientation control and eight of the twelve scientific instruments onboard.

The spacecraft's attitude control system relied on six hydrazine monopropellant thrusters, with two specifically dedicated to maintaining a constant spin rate of 4.8 revolutions per minute. A combination of star and Sun sensors allowed the spacecraft to track its orientation with Earth and other celestial objects, ensuring that Pioneer 11 remained on its planned trajectory throughout its mission.

Communication with Pioneer 11 was achieved through redundant transceivers attached to the high-gain and medium-gain antennas. Using the S-band frequency, the spacecraft transmitted data to Earth via the Deep Space Network. The spacecraft's limited onboard processing required ground operators to send pre-planned commands to be stored in memory for execution, ensuring precise control even at vast distances from Earth.

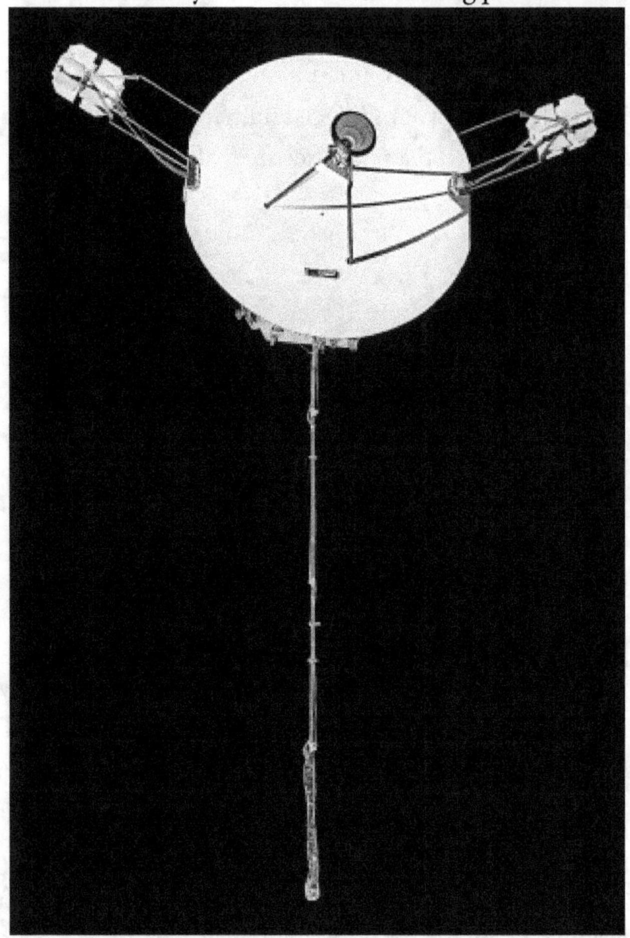

Four SNAP-19 radioisotope thermoelectric generators (RTGs) mounted on trusses extending three meters from the main body supplied power. At launch, the RTGs produced 155 watts of power, gradually decaying to 140 watts by the time the spacecraft reached Jupiter, more than enough to meet the spacecraft's 100-watt power requirement for its scientific operations.

Pioneer 11 was equipped with a sophisticated suite of scientific instruments designed to study the outer Solar System. Among its many tools, the spacecraft carried a helium vector magnetometer for mapping magnetic fields, a plasma analyzer to detect solar wind particles, and a cosmic ray telescope to study high-energy particles in space. It also boasted an ultraviolet photometer for measuring the quantities of hydrogen and helium in space, and an infrared radiometer for detecting cloud temperatures and heat emissions from both Jupiter and Saturn.

One of Pioneer 11's most significant contributions was its imaging photopolarimeter, which used the spacecraft's spin to create detailed images of Jupiter and Saturn in red and blue light. This instrument allowed scientists to study the planets' atmospheres and rings in unprecedented detail, providing crucial data for understanding the structure and composition of these gas giants.

In addition, Pioneer 11 carried an asteroid and meteoroid detector, which tracked particles in space, and a suite of instruments designed to study Saturn's ring system and atmospheric structure. The mission's data played a key role in planning the Voyager missions, which would later expand humanity's knowledge of the outer planets.

Pioneer 11 was launched on April 6, 1973, from Space Launch Complex 36A at Cape Canaveral, Florida, aboard an Atlas-Centaur rocket with a Star-37E propulsion module. Its launch followed that of its twin probe, Pioneer 10, by just over a year. Initially, Pioneer 11 was targeted for a direct trajectory to Jupiter. In May 1974, the spacecraft's trajectory was adjusted to facilitate a flyby of Saturn in 1979, a maneuver that increased its speed by 230 kilometers per hour and required 42 minutes and 36 seconds of thruster activity.

After making a series of course corrections, Pioneer 11 reached Jupiter in December 1974, flying within 42,000 kilometers of the planet's cloud tops. The spacecraft gathered invaluable data on Jupiter's magnetic field, radiation belts, and atmosphere, which would later inform the design of the Voyager spacecraft.

Pioneer 11's encounter with Saturn in September 1979 marked the first time a spacecraft had visited the ringed planet. It passed within 21,000 kilometers of Saturn, capturing detailed images of the planet's rings and moons, and providing critical data on the planet's magnetic field and atmosphere. The spacecraft's observations of Saturn's largest moon, Titan, offered tantalizing hints about its thick atmosphere, a mystery that the Cassini-Huygens mission would later explore in depth.

Pioneer 11's journey did not end with Saturn. The spacecraft continued on its path out of the Solar System, becoming the second artificial object, after Pioneer 10, to reach interstellar space. Although contact with the spacecraft was eventually lost, its contributions to space exploration remain invaluable. The mission paved the way for future exploration of the outer planets, laying the groundwork for the Voyager and Cassini missions, and expanding humanity's understanding of the vast and complex outer reaches of our Solar System.

In its final days, Pioneer 11's radioisotope thermoelectric generators continued to provide power, allowing the spacecraft to send its last data back to Earth, before fading into the vastness of interstellar space, a testament to human ingenuity and the relentless pursuit of knowledge beyond the stars.

Pioneer 11, launched on April 5, 1973, made its historic flyby of Jupiter in late 1974, following in the footsteps of its twin spacecraft, Pioneer 10. As the probe approached Jupiter, its mission objectives expanded to include detailed imaging of the planet's atmosphere, polar regions, and moons. On December 2, 1974, Pioneer 11 passed just 42,828 kilometers (26,612 miles) above Jupiter's cloud tops, capturing the first close-up images of the planet's Great Red Spot and unveiling new views of its polar regions.

During this flyby, Pioneer 11 gathered critical data on Jupiter's atmosphere and its moons, including Callisto, for which it provided an accurate measurement of mass. These findings contributed significantly to our understanding of Jupiter's immense gravitational influence. Leveraging this gravitational pull, Pioneer 11 performed a gravity assist maneuver that propelled it towards Saturn, accelerating its trajectory and altering its course. The spacecraft's velocity was increased, enabling the next stage of its journey to the outer Solar System. By April 16, 1975, after the successful Jupiter encounter, the micrometeoroid detector was deactivated, as it had completed its primary objective.

Pioneer 11's encounter with Saturn on September 1, 1979, was a milestone in space exploration, as it became the first spacecraft to reach the ringed giant. At its closest approach, the spacecraft flew within 21,000 kilometers (13,000 miles) of Saturn's cloud tops, providing unprecedented images and data about the planet's rings, moons, and atmosphere. This close pass was particularly significant because it confirmed the relative safety of flying through Saturn's ring plane, a critical factor for the Voyager missions that would follow.

Pioneer 11's mission was marked by high-risk, high-reward decisions. NASA's planners directed the spacecraft through Saturn's ring plane to test whether Voyager 1 and 2, following the same trajectory, could safely pass through without damage from potential ring particles. Acting as a true pioneer, the spacecraft's close encounter ensured that subsequent missions could explore Saturn and beyond without rerouting to avoid hazardous particles—a critical success, as avoiding the rings would have precluded the possibility of sending the Voyagers to Uranus and Neptune.

In a dramatic moment, Pioneer 11 nearly collided with one of Saturn's moons, passing just 4,000 kilometers (2,500 miles) from what was later identified as either Epimetheus or Janus. At the time of the encounter, these two moons were not fully understood to share the same orbit, leading to initial uncertainty about which moon had been photographed by Pioneer. The spacecraft also passed 103,000 kilometers (64,000 miles) from Mimas, another of Saturn's moons, providing additional data on Saturn's complex system of satellites.

The spacecraft's instruments detected another previously unknown moon and discovered a new ring, further contributing to our understanding of Saturn's vast and dynamic system. Detailed observations of Titan, Saturn's largest moon, revealed that it was too cold to support life as we know it, a key finding for future missions that would explore Titan's thick atmosphere in greater detail. Pioneer 11's images of Saturn's rings offered a novel perspective: the rings, typically bright when viewed from Earth, appeared dark in the spacecraft's photographs, while the gaps between them appeared bright.

After its successful encounters with Jupiter and Saturn, Pioneer 11 continued its journey out of the Solar System, entering the interstellar phase of its mission. On February 25, 1990, it became the fourth human-made object to pass beyond the orbit of the planets. Although its scientific instruments continued to function for many years, by 1995, the spacecraft's power levels had dropped too low to maintain full operations. NASA made the decision to terminate routine contact on September 30, 1995, after nearly 22 years of exploring the outer reaches of the Solar System.

In a press release marking the end of the mission, NASA Administrator Daniel Goldin praised Pioneer 11 as "the little spacecraft that could," highlighting its contributions to our understanding of the Solar System and human curiosity about the cosmos. Although routine operations ceased, NASA's Deep Space Network continued to listen for occasional signals from the spacecraft. The last meaningful data was received on November 24, 1995, marking the final chapter in Pioneer 11's journey.

As of June 2024, Pioneer 11 is approximately 113.1 astronomical units (AU) from Earth, traveling outward at a speed of 11.155 kilometers per second. It is headed in the direction of the constellation Scutum, near the star cluster Messier 26, and is expected to pass within 0.25 parsecs (0.82 light-years) of the star TYC 992-192-1 in about 928,000 years.

In the late 20th century, analysis of data from Pioneer 10 and 11 revealed a small but persistent anomaly in the spacecrafts' trajectories. Known as the "Pioneer anomaly," this unexplained drift in the Doppler frequency data suggested a constant acceleration towards the Sun. For years, scientists debated the source of this phenomenon, with hypotheses ranging from gravitational forces to new physics. However, after extended analysis by physicist Slava Turyshev and colleagues, the anomaly was attributed to asymmetric thermal radiation. The heat generated by the spacecraft's power systems created a small recoil force, gradually altering Pioneer's trajectory. This explanation resolved the mystery and confirmed that the Pioneer missions had, once again, provided valuable insights into both space exploration and fundamental physics.

Chapter 8 - The Shuttle Era and Expanding Frontiers

STS-1

First spaceplane in orbit, the Space Shuttle (test flight)

On April 12, 1981, the world witnessed the launch of STS-1 (Space Transportation System-1), the first orbital spaceflight of NASA's Space Shuttle program. This historic mission marked a significant leap in space exploration, as the Shuttle program promised reusable spacecraft for routine trips to low Earth orbit. The orbiter Columbia, the first of its kind, lifted off from Kennedy Space Center on a mission that would last 54.5 hours, orbiting the Earth 37 times before safely returning on April 14, 1981.

The mission carried a crew of two—commander John W. Young and pilot Robert L. Crippen. This flight was the first American crewed space mission since the Apollo-Soyuz Test Project (ASTP) in 1975, effectively ending a nearly six-year hiatus in American manned space exploration. STS-1 was not only Columbia's maiden flight but also the first test flight of the entire Space Shuttle system in space, though it had undergone extensive atmospheric testing (Approach and Landing Tests, ALT) and ground trials prior to launch.

In an interesting twist of history, the launch of STS-1 occurred on the 20th anniversary of Yuri Gagarin's pioneering flight aboard Vostok 1, the first human spaceflight. While the timing was purely coincidental—STS-1 had been delayed by two days due to technical issues—the connection between these two landmark events symbolized the ongoing legacy of human space exploration.

Commander John W. Young was already a veteran astronaut by the time of STS-1, with four prior missions under his belt, including two flights aboard Gemini and two on Apollo. He had even walked on the Moon in 1972 as the commander of Apollo 16. As the most experienced astronaut in NASA at the time, Young was the natural choice to command the first Space Shuttle flight, a decision he later admitted he made himself as Chief of the Astronaut Office. His vast experience made him the only remaining active member of NASA Astronaut Group 2, a prestigious group of astronauts selected in 1962.

Commander, John W. Young and Pilot Robert L.

Pilot Robert L. Crippen, on the other hand, was making his first trip into space, though his journey to the Shuttle program had been long and arduous. Crippen was part of NASA's Astronaut Group 7, selected after the cancellation of the U.S. Air Force's Manned Orbiting Laboratory (MOL) program. Prior to his selection for STS-1, Crippen participated in the Skylab Medical Experiment Altitude Test (SMEAT) and served as a capsule communicator (capcom) during the Skylab missions and the Apollo-Soyuz Test Project. STS-1 would make him the first astronaut from his group to fly in space.

In preparation for STS-1, both Young and Crippen underwent rigorous and unprecedented training, setting a record for the longest period any crew had trained for a mission before flying. Originally scheduled for launch in 1979, the mission was delayed several times, and the astronauts used the additional time to refine the Shuttle's systems, including helping design the cockpit's layout of 2,214 switches and displays—far more complex than Apollo's controls. Their work was critical in developing many contingency procedures, including a detailed 255-step process for handling an electronics failure due to a cooling system malfunction.

STS-1 was an engineering test flight, aimed at validating the Shuttle's performance in space. Columbia carried two Extravehicular Mobility Units (EMUs), spacesuits for potential emergency spacewalks. In the event of such a situation, Crippen would exit the orbiter with Young ready to assist if needed. Fortunately, no extravehicular activity was required.

Columbia launched with a mass of 99,453 kilograms (219,256 pounds), and returned slightly lighter at 88,662 kilograms (195,466 pounds), having jettisoned parts of its system during reentry. The mission's orbit had a perigee of 246 kilometers (153 miles) and an apogee of 274 kilometers (170 miles), with an orbital inclination of 40.30 degrees. The Shuttle orbited Earth in approximately 89.88 minutes per revolution.

In case of unforeseen issues, NASA had prepared a backup crew consisting of Commander Joe H. Engle and Pilot Richard H. Truly, both of whom would later fly on STS-2. The mission also had a dedicated support crew, including ascent capcom Daniel C. Brandenstein, Henry W. Hartsfield, and entry capcom Joseph P. Allen, all critical in ensuring smooth communication and operations during launch and landing.

During the initial planning stages for the Space Shuttle program, NASA, under the Carter Administration, faced the challenge of preparing for the Shuttle's first orbital flight. To mitigate potential risks, Vice President Walter F. Mondale, chairman of the National Space Council, suggested a suborbital flight that would land at an emergency landing site in Dakar, Senegal. NASA also proposed using the first Space Shuttle mission, STS-1, to test the Return to Launch Site (RTLS) abort scenario. This test involved an abort shortly after launch, where the Shuttle would jettison its solid rocket boosters (SRBs) and use its main engines to return to the launch site. The RTLS abort was a highly complex and dangerous maneuver, as it required precise timing and control to ensure the Shuttle's safe return.

John W. Young, a veteran astronaut who had flown to the Moon twice and walked on its surface, expressed strong reservations about these proposals. His experience and authority within NASA carried significant weight, leading to the eventual decision to forego both the suborbital flight and the RTLS abort test. Instead, STS-1 would proceed as an orbital mission. Young's straightforward concern summed up the risks of testing the RTLS abort: "Let's not practice Russian roulette, because you may have a loaded gun there."

Young's leadership and judgment set the stage for STS-1 to become the first fully crewed orbital flight of the Space Shuttle program.

STS-1 Mission Control Room in Houston

The maiden flight of the Space Shuttle, designated STS-1, launched on April 12, 1981, marking a monumental moment in space exploration. This historic date was not only the 20th anniversary of Yuri Gagarin's first human spaceflight but also the dawn of a new era in reusable spacecraft. The Space Shuttle orbiter Columbia lifted off from Pad A at Launch Complex 39 of the Kennedy Space Center at precisely 12:00:04 UTC.

The mission, commanded by John W. Young with Robert L. Crippen as the pilot, faced an early challenge. A launch attempt two days prior had been scrubbed due to a critical issue with Columbia's primary general-purpose computers, which failed to synchronize with the backup flight system. NASA engineers quickly implemented a software patch, resolving the issue in time for the successful launch on April 12.

STS-1 made history as the first U.S. crewed spaceflight to launch with solid-fuel rocket boosters, a departure from the liquid-fueled rockets used in previous crewed missions. Remarkably, the Shuttle was launched without an uncrewed test flight, further demonstrating NASA's confidence in its design. However, preparing Columbia for flight was no easy task; the orbiter spent an unprecedented 610 days in the Orbiter Processing Facility, largely due to the extensive replacement of heat shield tiles.

The mission's primary objective was straightforward but critical: to safely launch into orbit and return to Earth, verifying the Shuttle's spaceworthiness. The only payload on board was a Development Flight Instrumentation (DFI) package, which recorded data on the Shuttle's performance during all phases of the mission.

As Columbia's three RS-25 main engines roared to life, the noise was palpable, followed shortly by the ignition of the twin solid rocket boosters. Astronaut Robert Crippen described the Shuttle's lift-off as akin to being launched by a "steam catapult" from an aircraft carrier. The stack—the Shuttle, external tank, and boosters—began a smooth ascent, rolling to its planned azimuth of 67 degrees to achieve an orbital inclination of 40.3 degrees.

During the ascent, the Shuttle's engines throttled down to 65% to pass through Max Q, the point of maximum aerodynamic stress, before returning to full power. Just over two minutes into the flight, the SRBs separated, having propelled Columbia to an altitude of 53,000 meters—2,800 meters higher than planned. Columbia's main engines continued to fire for several more minutes until reaching Main Engine Cut-Off (MECO) at 8 minutes and 34 seconds Mission Elapsed Time (MET), achieving an altitude of 118,000 meters. The external tank was jettisoned 18 seconds later, eventually breaking up over the Indian Ocean.

Once in orbit, Young and Crippen set to work, deploying the payload bay doors, which allowed Columbia's radiators to dissipate heat from onboard systems. Opening the doors was critical; failure to do so would have necessitated an early return to Earth. The astronauts discovered minor damage to the thermal protection system (TPS) tiles on the Orbital Maneuvering System (OMS) pods, a concern that would become a recurring theme in Shuttle missions.

The majority of the 54-hour mission was spent conducting systems tests, ensuring that all components of the Shuttle functioned as expected. Tests included star tracker performance, RCS testing, and fuel cell purging. Despite the challenges of scheduling external imaging of Columbia's heat shield, all 113 flight test objectives were completed successfully.

Columbia's cargo bay and aft section, on 12 April 1981.

Notably, Vice President George H.W. Bush called the astronauts during their mission, a gesture of support from the White House. President Ronald Reagan, still recovering from an assassination attempt, had intended to visit Mission Control but was unable to do so.

Preparations for reentry began with the closure of the payload bay doors. The de-orbit burn took place during the 36th orbit of the mission, over the southern Indian Ocean. Columbia reentered the atmosphere at an altitude of approximately 120,000 meters, beginning a 21-minute communications blackout as it streaked through the atmosphere at hypersonic speeds.

As Columbia crossed the California coast, Young took manual control for the final phase of the descent. The Shuttle touched down on the dry lakebed at Edwards Air Force Base at 18:21 UTC on April 14, 1981, at a speed of 339 kilometers per hour. The successful landing concluded a 36-orbit mission that lasted 2 days, 6 hours, and 20

Columbia landing on Rogers Dry Lake bed at Edwards
Air Force Base, on 14 April 1981.

minutes, covering over 1.7 million kilometers.

After landing, Young proclaimed, "This is the world's greatest all-electric flying machine. That was super!" The Shuttle program had passed its first critical test, marking the beginning of a new era in space exploration.

STS-1, the first orbital test flight of the Space Shuttle, was one of the most ambitious and complex space missions undertaken by NASA at the time. The Shuttle Columbia was hailed as the most intricate flying machine ever built, incorporating cutting-edge technology across multiple systems. Despite this, the mission faced numerous technical challenges, with approximately 70 anomalies observed during and after the flight. These issues highlighted the difficulties of testing a spacecraft as complex as the Shuttle in its inaugural flight. Nevertheless, Columbia completed the mission successfully, marking a significant milestone in space exploration. Below are some of the key anomalies that occurred during STS-1.

Columbia, mated to the Shuttle Carrier Aircraft, arrives at Kennedy Space Center after STS-1 to be prepared for its next mission.

Engineers, much like during the first Saturn V launch in 1967, underestimated the noise and vibration levels produced by the Shuttle's solid rocket boosters (SRBs). The intense thrust generated by the SRBs created shock waves that were deflected into Columbia's tail section, causing its wing flaps to flex and bending several fuel tank supports. These unexpected structural stresses could have made landing difficult, especially if the flaps had been significantly damaged. To address this issue, NASA later installed an improved sound suppression system at Launch Complex 39A to mitigate the vibrations during future launches.

During the ascent, pilot Robert Crippen reported seeing "white stuff" splattering the Shuttle's windows. This material was likely the white paint covering the external tank's thermal foam, which peeled off during launch. Although the debris did not pose a significant risk to the mission, it raised concerns about potential damage to the Shuttle's delicate thermal protection system .

One of the more serious anomalies involved damage to the Shuttle's thermal protection tiles, particularly on the Orbital Maneuvering System (OMS) and Reaction Control System (RCS) pods at the aft end of the orbiter. Upon visual inspection during orbit, Commander John Young reported that two tiles on the Shuttle's nose appeared as though "big bites" had been taken out of them .

In a precautionary measure, the U.S. Air Force used a KH-11 reconnaissance satellite to photograph Columbia's underside to assess the extent of the damage. Due to the classified nature of the satellite, only a few NASA personnel were aware of the arrangement. Young and Crippen were instructed to perform specific maneuvers using the RCS thrusters to enable the satellite to capture images, though they were not informed of the reason behind the maneuvers. The photos confirmed that the damage was not critical, but approximately 16 undensified tiles near the OMS pod had been lost during ascent .

During reentry, Columbia's aerodynamics at high Mach numbers diverged from pre-flight predictions. The actual location of the center of pressure differed from estimates, forcing the Shuttle's flight computer to extend the body flap by 16 degrees instead of the anticipated 8 or 9 degrees. Additionally, the first roll maneuver during reentry caused lateral and directional oscillations, resulting in twice as high side slip angles as expected. This anomaly was attributed to larger-than-anticipated rolling moments caused by yaw RCS jet firings .

The Shuttle's heat shield sustained damage when an overpressure wave from the SRBs caused a forward RCS oxidizer strut to fail. This same overpressure wave also forced the body flap—a critical component that controls pitch during reentry—beyond its designed range of motion. Had the body flap's hydraulic system ruptured, the Shuttle would have been unable to maintain control during descent, making a controlled landing impossible. John Young later revealed that, had the crew been aware of this damage, they would have attempted to fly to a safe altitude and eject, abandoning the Shuttle .

The extreme heat encountered during reentry melted and distorted the strike plate next to Columbia's external tank door. An improperly installed thermal protection tile near the plate caused the damage. Though it did not impact the mission's success, it was a point of concern for future flights.

One of the most serious, and previously undisclosed, issues involved a protruding tile gap filler that ducted hot gas into the right main landing gear compartment during reentry. This caused significant damage, including the buckling of the landing gear door. John Young mentioned that neither he nor Crippen had been informed of this issue until reading the post-flight mission report, which documented the gas leak but not the door buckling. The anomaly report, however, did note the door damage (STS-1-V-49)

Despite these numerous anomalies, the STS-1 mission was completed successfully, and Columbia's overall performance was commendable. The data gathered from the mission allowed NASA to make crucial modifications to both the Shuttle and its launch and reentry procedures. This included installing better sound suppression systems, revising thermal protection protocols, and improving the Shuttle's structural resilience. Columbia went on to fly four more Shuttle missions, demonstrating NASA's commitment to learning from each flight and continuously improving the Shuttle program .

The lessons learned from STS-1 set the foundation for the future of reusable spacecraft, ensuring that subsequent missions would benefit from the extensive testing and analysis conducted during the Shuttle's first orbital flight.

The launch of STS-1 on April 12, 1981, held deep historical significance, coinciding with the 20th anniversary of Yuri Gagarin's groundbreaking Vostok 1 mission, which carried the first human into space. This shared anniversary linked two monumental achievements in space exploration. In recognition of both events, Yuri's Night was established in 2001 as an annual celebration of space exploration, paying tribute to Gagarin's flight and the first launch of the Space Shuttle.

In honor of the 25th anniversary of STS-1, NASA renamed Firing Room 1 at the Launch Control Center in Kennedy Space Center—the control room where Columbia's first flight was launched—as the Young-Crippen Firing Room. This commemorated the daring mission led by Commander John W. Young and Pilot Robert L. Crippen. NASA described the mission as "the boldest test flight in history," a testament to the risks and challenges faced by the crew and the Space Shuttle program.

STS-1 and the subsequent STS-2 mission were unique in that the Shuttle's external fuel tank was painted white. NASA opted to paint the tank as an additional protective measure, though this decision was short-lived. Beginning with STS-3, the tanks were left unpainted, revealing the characteristic orange hue of the insulation foam. This change, while subtle, provided a weight savings of approximately 272 kilograms (600 pounds). Over time, the orange color became an iconic visual hallmark of the Space Shuttle, distinguishing it from earlier missions and highlighting NASA's continuous efforts to optimize the Shuttle for efficiency and performance.

While STS-1 ultimately achieved great success, its preparation was marred by a tragic accident that claimed the lives of three technicians: John Bjornstad, Forrest Cole, and Nicholas Mullon. On March 19, 1981, during a countdown demonstration test, a nitrogen purge was introduced into the aft engine compartment of Columbia to reduce the risk of explosion from other volatile gases present in the orbiter. However, a procedural oversight allowed workers to re-enter the compartment before it was fully purged of nitrogen, a gas that is both colorless and odorless.

The three technicians, unaware of the hazard, entered without air packs, leading to their unconsciousness due to oxygen deprivation.

A fourth worker, upon discovering the men, attempted a rescue but was also incapacitated. It was not until two additional workers arrived that help was finally summoned. Although security guards with air packs managed to remove the unconscious men from the compartment, the delay in medical response proved fatal. Bjornstad died on site, Cole succumbed to his injuries on April 1 without regaining consciousness, and Mullon, having sustained permanent brain damage, died in 1995 due to complications from the accident.

This incident was the first fatality at Cape Canaveral's launchpad since the Apollo 1 disaster in 1967, which claimed the lives of three astronauts during a pre-flight test. Despite the loss, the launch of STS-1 proceeded as scheduled less than a month later. While in orbit, Robert Crippen delivered an emotional tribute to Bjornstad and Cole, acknowledging their dedication to the space program and the significance of their contributions.

A subsequent three-month investigation determined that the accident was caused by a combination of recent procedural changes and communication breakdowns. The findings were documented in the "LC-39A Mishap Investigation Board Final Report." The names of the three men—John Bjornstad, Forrest Cole, and Nicholas Mullon—are engraved on a monument at the U.S. Space Walk of Fame in Florida, ensuring their legacy within the history of space exploration.

Though this tragedy cast a shadow over the preparations for STS-1, the mission's success marked the beginning of a new chapter in human spaceflight, forever changing the way NASA—and the world—looked at space travel.

STS-41-B

First untethered spacewalk, Bruce McCandless II

STS-41-B: A Milestone Mission in Space Shuttle History

STS-41-B was NASA's tenth Space Shuttle mission and the fourth flight of the Space Shuttle Challenger, marking significant milestones in the Space Shuttle program. Launched on February 3, 1984, and concluded with a successful landing on February 11, 1984, this mission was notable for the deployment of two communications satellites and the first untethered spacewalk, a landmark event in human spaceflight.

Vance D. Brand, commander; and Robert L. Gibson, pilot. Standing left to right are mission specialists Robert L. Stewart, Ronald E. McNair, and Bruce McCandless.

STS-41-B was initially designated as STS-11, but the flight numbering system changed following the STS-9 mission. The next mission, STS-10, was canceled due to payload delays, leading STS-11 to be reclassified as STS-41-B. This shift was part of NASA's transition to a new system that reflected both the year of the mission and the launch site.

The STS-41-B crew consisted of five astronauts, led by Commander Vance D. Brand, who was making his second Shuttle flight. Joining him were Pilot Robert L. Gibson and Mission Specialists Bruce McCandless II, Ronald E. McNair, and Robert L. Stewart, all flying for the first time. Their teamwork and expertise were essential in achieving the mission's objectives, including satellite deployment, spacewalks, and scientific experiments.

Challenger lifted off from Kennedy Space Center at 8:00 a.m. EST on February 3, 1984, with an estimated 100,000 spectators in attendance. Approximately eight hours into the mission, the crew deployed two communications satellites—Westar 6 for Western Union and Palapa B2 for Indonesia—both built by Hughes Aircraft as part of their HS-376 series. However, the Payload Assist Modules (PAM) responsible for boosting the satellites into their intended orbits malfunctioned, leaving them in lower-than-planned orbits. These satellites were later retrieved successfully by the crew of STS-51-A in November 1984, aboard Space Shuttle Discovery.

On February 7, 1984, Bruce McCandless II and Robert L. Stewart conducted the first untethered spacewalk in history, an extraordinary moment for space exploration. Using the Manned Maneuvering Unit (MMU), McCandless ventured 98 meters (322 feet) from the shuttle, maneuvering freely in space for the first time. Stewart tested a foot restraint attached to the Remote Manipulator System (Canadarm), practicing procedures for the upcoming Solar Maximum Mission (SMM) satellite retrieval and repair on the following mission, STS-41-C. This landmark extravehicular activity (EVA) lasted 5 hours and 55 minutes, and a second EVA followed on February 9, lasting 6 hours and 17 minutes.

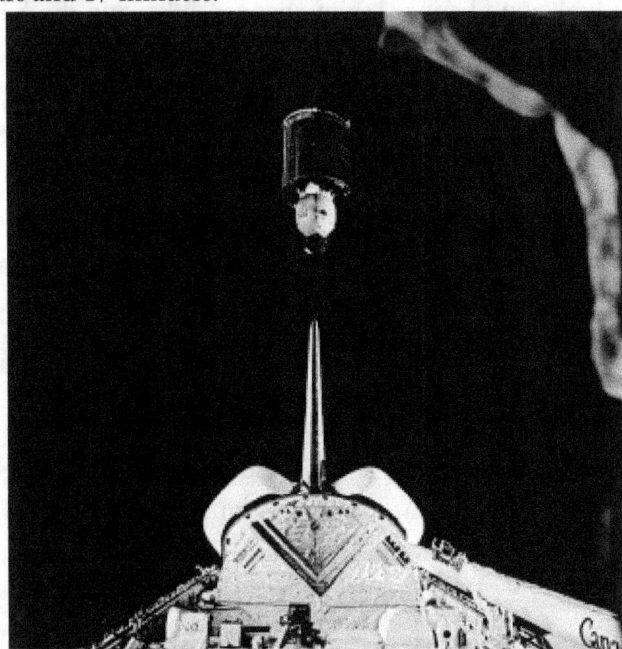

Palapa B-2 satellite after deployment on STS-41-B. View of the Palapa-B and the Shuttle Challenger begining their separation after deployment of the communications satellite.

STS-41-B carried a variety of scientific experiments that expanded the boundaries of research in microgravity. Among them was the reflight of the West German-sponsored SPAS-1 pallet, first flown on STS-7, although it remained in the payload bay due to an electrical problem with the Canadarm. The mission also included five Get Away Special (GAS) canisters, six live rats as part of biological experiments, a Cinema-360 camera for filming in space, and ongoing studies with the Continuous Flow Electrophoresis System and the Monodisperse Latex Reactor.

A highlight of the scientific payload was the first student-designed experiment to fly in space. Developed by a team from Brighton High School in Utah, in partnership with Utah State University, the experiment investigated seed germination and plant growth in the weightless space environment. This initiative underscored NASA's commitment to inspiring the next generation of scientists and engineers.

Despite the successes, STS-41-B faced technical issues that tested the resilience of both the shuttle and the crew. The Challenger's supply and wastewater venting systems encountered below-freezing temperatures, causing the water dump valve to fail. Excess water was diverted through the flash evaporator for the remainder of the mission. Additionally, during re-entry, ice formed on the dump valves, and pieces broke off, striking the left Orbital Maneuvering System (OMS) pod and damaging several Thermal Protection System (TPS) tiles. Although minimal damage did not compromise the shuttle's safety, a post-flight inspection revealed ruptured lines and missing insulation around the nozzles.

After completing 128 orbits and traveling over 5.3 million kilometers (3.3 million miles), Challenger landed at Kennedy Space Center's Shuttle Landing Facility on February 11, 1984, marking the first time a spacecraft landed at its launch site. The mission lasted 7 days, 23 hours, 15 minutes, and 55 seconds, cementing its place in the annals of space exploration history.

STS-41-B was a significant mission, demonstrating NASA's capabilities in satellite deployment, extravehicular activity, and the safe return of both crew and orbiter. The first untethered spacewalk and the various scientific contributions made during this flight laid the groundwork for future advances in space exploration, including the repair of satellites and further study of the effects of microgravity on living organisms.

Vega 1 probe

First aerostat balloon in the atmosphere of Venus

Vega 1, along with its twin Vega 2, was part of the Soviet Union's ambitious Vega program, aimed at exploring both Venus and Halley's Comet. This dual mission was an evolution of the earlier Venera spacecraft, developed by the Babakin Space Centre and constructed as the 5VK model by Lavochkin in Khimki. The name "Vega" (Вега) was derived from the Russian words for Venus (Венера: "Venera") and Halley (Галлея: "Galleya"), reflecting the mission's two-fold objective.

The spacecraft, weighing 4,840 kilograms (10,670 lbs), was powered by two large solar panels and equipped with a suite of scientific instruments. These included a high-gain antenna, cameras, a spectrometer, an infrared sounder, magnetometers (MISCHA), and plasma probes. Both Vega 1 and Vega 2 were launched by Proton-K rockets from the Baikonur Cosmodrome in June 1985. Each was a three-axis stabilized spacecraft, built with a dual bumper shield to protect from dust particles during the comet encounter.

Vega 1's descent module separated from the main spacecraft on June 9, 1985, and arrived at Venus two days later, on June 11, 1985. The descent module, weighing 1,500 kilograms (3,300 lbs) and measuring 240 centimeters (94 inches) in diameter, carried a surface lander and a balloon explorer. As the flyby probe performed a gravitational assist to continue toward Halley's Comet, the descent module studied the Venusian atmosphere and surface.

The surface lander, similar to those used in the Venera program, carried an array of scientific tools including temperature and pressure sensors, an ultraviolet spectrometer, a gas chromatograph, an X-ray spectrometer, and a surface sampling device. Its primary objective was to study the dense Venusian atmosphere and surface composition. However, due to extreme turbulence during descent, some experiments were prematurely activated 20 kilometers above the surface, and only the mass spectrometer was able to return data before the lander reached the surface at 7.2°N, 177.8°E, in the Mermaid Plain near Aphrodite Terra. Unfortunately, since the landing occurred during nighttime, no surface images were captured.

One of the most innovative aspects of the Vega mission was the use of a balloon probe to study the Venusian atmosphere at an altitude where conditions were less hostile. After entering the Venusian atmosphere at approximately 11 kilometers per second (6.8 miles per second), the balloon was deployed at around 54 kilometers (34 miles) altitude. It quickly inflated and began drifting westward in the planet's strong zonal winds at speeds of up to 69 meters per second (154 mph). The balloon floated at a stable altitude of 53.6 kilometers (33.3 miles), where atmospheric pressure measured 535 millibars, and temperatures ranged between 27 to 37°C (80 to 98°F).

The balloon probe crossed the night-to-day terminator on June 12, 1985, after traveling 8,500 kilometers (5,300 miles). Its final transmission was received on June 13 after it had traveled a total of 11,600 kilometers (7,200 miles), covering nearly 30% of the planet's circumference. It is unclear how much farther the balloon traveled after losing contact, but the data gathered provided unprecedented insights into the dynamic Venusian atmosphere.

Following the Venus flyby, Vega 1 used a gravity assist from the planet to continue toward its second target: Halley's Comet. The spacecraft made its closest approach to the comet on March 6, 1986, at 8,889 kilometers (5,523 miles) from the nucleus. During this encounter, Vega 1 captured over 500 images using various filters, helping scientists refine the trajectory for the European Space Agency's Giotto spacecraft, which would make an even closer flyby of the comet.

Vega 1's images revealed two bright jets emanating from the nucleus, initially mistaken for a double nucleus. These jets indicated active outgassing from the comet's icy core. Further analysis of the data showed the nucleus to be darker than expected, with a temperature ranging between 300 and 400 K (27 to 127°C), much warmer than anticipated for an icy body. The spacecraft's instruments also measured the nucleus to be approximately 14 kilometers (8.7 miles) long, with a rotation period of about 53 hours.

The dust mass spectrometer on Vega 1 detected materials consistent with carbonaceous chondrite meteorites, and the spacecraft's observations contributed significantly to our understanding of the comet's composition and the interaction between the solar wind and the comet's gas and dust.

After the intense examination of Halley's Comet, which lasted about three hours during closest approach, Vega 1 continued its mission in deep space, sending back additional images on March 7 and 8, 1986. Vega 1 and Vega 2 returned over 1,500 images of Halley's Comet. However, by January 30, 1987, Vega 1 had exhausted its attitude control propellant, and communication ceased. Its twin, Vega 2, remained operational until March 24, 1987.

Both spacecraft are now in heliocentric orbits. Vega 1 orbits the Sun with a perihelion of 0.70 AU, an aphelion of 0.98 AU, and an orbital period of 281 days. Its remarkable dual mission, exploring both Venus and Halley's Comet, remains one of the most ambitious and scientifically productive achievements of the Soviet space program.

Voyager 2

First Uranus flyby
First Neptune flyby

Voyager 2: A Journey Through the Outer Solar System

Launched by NASA on August 20, 1977, Voyager 2 was a pioneering space probe, part of the ambitious Voyager program, designed to explore the outer planets of our solar system. The spacecraft was launched on a trajectory that would take it past the gas giants Jupiter and Saturn, and further towards the distant ice giants Uranus and Neptune. To this day, Voyager 2 remains the only spacecraft to have visited Uranus and Neptune, providing humanity with unprecedented insights into these distant worlds. It was the third of five spacecraft to achieve solar escape velocity, a feat allowing it to leave the solar system and continue its journey into interstellar space. As of June 2024, Voyager 2 has been in operation for over 47 years, making it the oldest active space probe in human history.

Voyager 2's mission began 16 days before its twin, Voyager 1, which followed a different trajectory. While Voyager 2's primary mission was to study the outer planets, its extended mission involves the study of the interstellar medium beyond the heliosphere—the vast bubble of charged particles generated by the Sun's influence.

The concept behind the Voyager missions originated in the early days of space exploration. During the 1970s, NASA identified a rare planetary alignment that would enable a single spacecraft to visit multiple planets by using gravity assists—a method whereby a spacecraft uses a planet's gravity to slingshot toward the next destination. This alignment, which occurs only once every 175 years, would allow a probe to visit Jupiter, Saturn, Uranus, and Neptune in a single mission.

Originally conceived as part of the "Grand Tour" program, Voyager 2 was one of two probes developed to take advantage of this opportunity. Early plans called for even more ambitious missions, involving separate probes to visit Pluto, but budget constraints led to a more modest plan. By 1972, the mission was scaled back, and the Voyager spacecraft were reconfigured as successors to the earlier Mariner probes, focusing initially on flybys of Jupiter and Saturn. Nonetheless, the possibility of extending the mission to Uranus and Neptune remained open, should conditions permit.

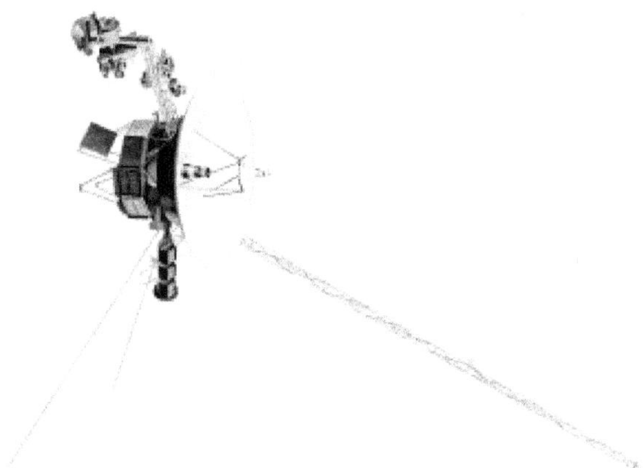

After Voyager 1 successfully completed its exploration of Jupiter, Saturn, and the largest of Saturn's moons, Titan, Voyager 2 was granted an extension to continue its journey to Uranus and Neptune. These missions, which took place between 1979 and 1989, provided the first and only close-up images of these distant ice giants.

Voyager 2 first encountered Jupiter in 1979, capturing detailed images of the planet's cloud bands and discovering active volcanic activity on its moon, Io. In 1981, the spacecraft reached Saturn, transmitting data about the planet's ring system and its numerous moons. The next leg of the journey was more challenging, as Voyager 2 set course for Uranus, a planet previously unseen in close detail. In 1986, Voyager 2 flew past Uranus, revealing its unique tilt and detecting 10 previously undiscovered moons. Continuing its voyage, Voyager 2 made its closest approach to Neptune in 1989, capturing stunning images of the planet's blue atmosphere and its largest moon, Triton, which was found to be geologically active with geysers erupting nitrogen gas.

By the conclusion of this grand tour, Voyager 2 had fundamentally changed humanity's understanding of the outer planets and their moons. Each flyby revealed new surprises, offering insights into these distant worlds' dynamic and diverse environments.

Following its encounter with Neptune, Voyager 2 began its extended mission to study the outermost regions of the solar system and beyond. On November 5, 2018, at 119.7 astronomical units (AU) from the Sun, Voyager 2 crossed the boundary of the heliosphere and entered interstellar space. This region, known as the interstellar medium, is composed of plasma and particles that exist between the stars. It joined its twin, Voyager 1, which had crossed into interstellar space six years earlier in 2012. Now beyond the Sun's influence, Voyager 2 continues to send back valuable scientific data, including the first direct measurements of the density and temperature of interstellar plasma.

Built by NASA's Jet Propulsion Laboratory (JPL), Voyager 2 was designed with redundancy and durability in mind, ensuring it could survive its extended journey far from Earth. The spacecraft's high-gain antenna, measuring 3.7 meters (12 feet) in diameter, continues to communicate with Earth via the Deep Space Network, relaying scientific data across billions of kilometers. Voyager 2 is equipped with 16 hydrazine thrusters for attitude control, and its power is generated by three radioisotope thermoelectric generators (RTGs), which convert heat from decaying plutonium into electricity. These RTGs provided 470 watts of power at launch, enough to keep the probe's scientific instruments operational.

Over the years, the power available to the spacecraft has steadily diminished, halving approximately every 87 years. By 2023, Voyager 2's power levels were declining, and NASA began tapping into a backup power reserve intended for an onboard safety mechanism. This move has enabled the spacecraft to continue operating its five remaining scientific instruments through at least 2026.

The suite of scientific instruments aboard Voyager 2 includes a variety of sensors to study the magnetic fields, plasma, and charged particles in the outer planets and interstellar space. Among these are the Triaxial Fluxgate

Magnetometer (MAG), which investigates magnetic fields; the Plasma Spectrometer (PLS), which measures electrons and plasma ions; and the Cosmic Ray System (CRS), designed to study high-energy cosmic rays. Instruments like the Infrared Interferometer Spectrometer (IRIS) and the Ultraviolet Spectrometer (UVS) were essential for studying the atmospheres of the outer planets, although some instruments, such as the imaging systems, are now disabled.

Despite these limitations, Voyager 2 continues to provide critical data as it travels farther from Earth. As of June 2024, the probe is located 136.1 AU (20.4 billion kilometers) from our planet, moving through the vast expanse of interstellar space at a speed of 15.34 kilometers per second (34,320 mph).

Voyager 2 Timeline of Travel

August 20, 1977: Voyager 2 was launched from Cape Canaveral, Florida, at 14:29 UTC, aboard a Titan IIIE/Centaur launch vehicle. This marked the beginning of its historic journey to the outer planets and beyond.

December 10, 1977: The spacecraft entered the asteroid belt, a region filled with rocky debris orbiting the Sun between Mars and Jupiter.

December 19, 1977: Voyager 1 overtook Voyager 2 despite being launched later on September 5, 1977, due to its shorter trajectory.

June 1978: The spacecraft's primary radio receiver failed, and the mission continued using its backup radio receiver for the remainder of the journey.

October 21, 1978: Voyager 2 exited the asteroid belt, continuing its journey toward Jupiter.

April 25, 1979: The spacecraft began its observation phase of Jupiter, gathering detailed images and data about the gas giant and its moons.

June 5, 1981: Voyager 2 started its observation phase of Saturn, capturing images of the planet's famous rings and moons.

November 4, 1985: The spacecraft began observing Uranus, providing humanity's first close-up views of the ice giant and its moons.

August 20, 1987: Voyager 2 celebrated 10 years of continuous flight and operation, having traveled vast distances across the solar system.

June 5, 1989: The spacecraft entered the observation phase of Neptune, the final planet in its primary mission. It provided detailed images of the blue planet and its moon Triton.

October 2, 1989: The Voyager Interstellar Mission (VIM) began after completing its exploration of Neptune, transitioning Voyager 2 to its new mission of studying the outer reaches of the solar system and the interstellar medium.

Interstellar Phase

August 20, 1997: The spacecraft reached 20 years of continuous flight and operation, still transmitting data from deep space.

November 13, 1998: NASA terminated the scan platform and ultraviolet (UV) observations to conserve power as the spacecraft continued its journey toward interstellar space.

August 20, 2007: Voyager 2 achieved 30 years of continuous flight and operation.

September 6, 2007: The spacecraft's data tape recorder operations were terminated, further conserving power.

February 22, 2008: Operations for the planetary radio astronomy experiment were terminated.

November 7, 2011: NASA switched to Voyager 2's backup thrusters to conserve power, ensuring that the spacecraft could continue its journey as long as possible.

August 20, 2017: Voyager 2 reached its 40th anniversary of continuous flight and operation, continuing to transmit scientific data from the farthest reaches of the solar system.

November 5, 2018: Voyager 2 crossed the heliopause, the boundary where the solar wind from the Sun meets the interstellar medium, marking its official entry into interstellar space. This monumental achievement followed Voyager 1, which had crossed the same boundary six years earlier.

July 18, 2023: Voyager 2 overtook Pioneer 10 to become the second farthest spacecraft from the Sun, continuing its mission in the vast expanse of interstellar space.

Launch and Trajectory

Launched on August 20, 1977, from Space Launch Complex 41 at Cape Canaveral, Florida, Voyager 2 embarked on its epic journey aboard a Titan IIIE/Centaur launch vehicle. Two weeks later, its twin, Voyager 1, was launched on September 5, 1977. While Voyager 1 followed a shorter, faster trajectory to Jupiter and Saturn, Voyager 2 traveled on a more circular orbit.

Voyager 1 reached Jupiter and Saturn earlier due to its trajectory, which had an aphelion (the farthest point from the Sun) of 8.9 AU—just short of Saturn's orbit. In contrast, Voyager 2's initial orbit had an aphelion of 6.2 AU, well within Saturn's orbital path. Despite this longer journey, Voyager 2 was set to explore Uranus and Neptune, which Voyager 1 would not visit.

In April 1978, a critical communications issue arose when no commands were sent to Voyager 2 for a brief period, forcing the spacecraft to switch from its primary radio receiver to the backup. Later, the primary receiver failed completely. Although the backup was functional, it had a malfunctioning capacitor, which required NASA to transmit signals at a precise frequency. This precision was necessary because the receiver was sensitive to Earth's rotation (due to the Doppler effect) and the onboard temperature.

Despite these challenges, Voyager 2 continued its journey, collecting valuable data and transmitting it back to Earth via the Deep Space Network.

On July 9, 1979, Voyager 2 made its closest approach to Jupiter, coming within 570,000 km (350,000 mi) of the gas giant's swirling cloud tops. At 22:29 UTC, the spacecraft began transmitting some of the most detailed images of Jupiter ever seen, revealing a wealth of new information about the planet's atmosphere, moons, and dynamic weather systems. One of the mission's most striking discoveries was the complexity of Jupiter's Great Red Spot, a massive storm moving counterclockwise, driven by immense winds. In addition to the Great Red Spot, Voyager 2 uncovered smaller storms and eddies scattered throughout the planet's cloud bands, painting a dynamic picture of Jupiter's atmosphere.

During this encounter, Voyager 2 also captured images of Jupiter's moons—Amalthea, Io, Callisto, Ganymede, and Europa. Of particular interest was Io, where Voyager 2 confirmed Voyager 1's observations of active volcanism. In a 10-hour "volcano watch," Voyager 2 detected volcanic eruptions on Io, showing that the moon's surface had undergone significant changes since Voyager 1's visit just four months earlier. Both Voyager spacecraft observed nine eruptions, with additional evidence suggesting that more occurred between the two flybys.

Europa, one of Jupiter's most intriguing moons, displayed intersecting linear features, initially thought to be deep cracks caused by tectonic activity. Closer inspection by Voyager 2 revealed these features had no significant topography, puzzling scientists. This discovery led to the hypothesis that Europa's icy crust, possibly less than 30 km (19 mi) thick, could be floating atop a 50 km (31 mi)-deep ocean, with tidal heating driving the moon's internal activity.

Voyager 2 also made significant discoveries in Jupiter's rings and smaller moons. It identified two new moons, Adrastea and Metis, orbiting just outside Jupiter's rings, and a third new moon, Thebe, located between the orbits of Amalthea and Io.

On August 26, 1981, Voyager 2 made its closest approach to Saturn, passing behind the planet as viewed from Earth. This positioning allowed the spacecraft to use its radio link to study Saturn's upper atmosphere. Voyager 2 recorded temperatures and pressures at different depths of Saturn's atmosphere, finding that the upper layers were cooler than expected, with temperatures around 82 K (−191°C). As it descended deeper into the atmosphere, where pressures were higher, the temperature increased to 143 K (−130°C). The data collected by Voyager 2 indicated that Saturn's north pole was about 10°C cooler than its mid-latitudes, a temperature difference likely caused by seasonal variations.

However, the Saturn encounter was not without complications. After its flyby, Voyager 2 experienced a malfunction in its scan platform, where the azimuth actuator seized, affecting its ability to gather data. Engineers traced the issue to a combination of design flaws, including the lubrication system and corrosion. Despite this, the mission team successfully sent commands that partially restored the scan platform's function, allowing Voyager 2 to continue its journey.

Voyager 2's Saturn flyby offered a wealth of scientific discoveries. It imaged Saturn's moons, observed spoke-like features in Saturn's rings, and studied the atmosphere of Titan, Saturn's largest moon, from 2.3 million km. While Voyager 2 did not perform a close flyby of Titan, it provided valuable data on Titan's thick, nitrogen-rich atmosphere, which was later explored in detail by subsequent missions.

On January 24, 1986, Voyager 2 made its closest approach to Uranus, passing within 81,500 km (50,600 mi) of the planet's cloud tops. This marked the first and only visit to Uranus by a spacecraft, providing a groundbreaking look at the planet's atmosphere, rings, and moons. Voyager 2 discovered 11 previously unknown moons orbiting Uranus, including Cordelia, Ophelia, Bianca, and Juliet, among others.

One of the most significant findings from the Uranus encounter was the planet's unusual magnetic field, which is tilted by nearly 60 degrees relative to its rotational axis. This misalignment created a twisted magnetic tail extending millions of kilometers into space. The spacecraft also detected lightning-like electrostatic discharges, further evidence of the planet's dynamic atmospheric processes.

Though much of Uranus' cloud features were hidden beneath a thick haze, Voyager 2 captured false-color images showing bands of clouds encircling the planet's south pole. The planet's average atmospheric temperature was found to be about 60 K (−213°C), making Uranus one of the coldest planets in the solar system. Despite this frigid environment, Voyager 2 discovered dayglow emissions in ultraviolet light at the planet's south pole, hinting at complex interactions between Uranus' atmosphere and magnetic field.

The spacecraft also revealed previously unseen features of Uranus' moons, particularly Miranda, which exhibited massive canyons and cliffs, suggesting a history of intense geological activity. One hypothesis proposed that Miranda had been shattered and reassembled multiple times, resulting in its fractured surface.

On August 25, 1989, Voyager 2 made its closest approach to Neptune, the last of the outer planets in its grand tour. This encounter required precise navigation due to the tilt of Neptune's largest moon, Triton, relative to the planet's orbit. Flight controllers at NASA carefully plotted a trajectory that allowed Voyager 2 to pass within 4,950 km (3,080 mi) of Neptune's north pole and five hours later, make a close flyby of Triton, coming within 40,000 km (25,000 mi) of the moon.

Voyager 2's observations revealed Neptune's "Great Dark Spot," a storm similar to Jupiter's Great Red Spot, but much shorter-lived. This storm, thought to be a window in Neptune's methane cloud deck, later disappeared, as confirmed by observations from the Hubble Space Telescope. Neptune's atmosphere also displayed high winds and cold temperatures, averaging around 60 K (−213°C).

The flyby of Triton revealed active geysers erupting nitrogen gas, suggesting that Triton, despite its frigid surface, was geologically active. The spacecraft also discovered six new moons, including Despina, Galatea, and Proteus, as well as additional rings around Neptune, previously unseen by Earth-based telescopes.

In 2020, scientists reevaluating Voyager 2's data from the Neptune encounter discovered a large atmospheric magnetic bubble, or plasmoid, released from the planet. This finding highlighted the ongoing potential of Voyager 2's data to reveal new insights, even decades after the spacecraft's historic flybys.

Voyager 2's Interstellar Mission: A Journey Beyond the Solar System

On November 5, 2018, Voyager 2 crossed a monumental boundary—it left the heliosphere, entering the vast, uncharted territory of interstellar space. This transition marked the spacecraft's shift from its planetary mission to its interstellar mission, a journey that continues to provide valuable insights into the space beyond our solar system. By September 2023, Voyager 2 was still transmitting scientific data back to Earth, albeit at a much-reduced rate of 160 bits per second. Despite the distance and technological challenges, the spacecraft continues to contribute to humanity's understanding of the universe beyond the Sun's influence.

The Interstellar Mission's Objectives

The primary objective of Voyager 2's interstellar mission is to study the region beyond the heliosphere—the vast bubble of solar wind and magnetic fields that surrounds our solar system. As the probe moved into the interstellar medium, it began transmitting the first direct measurements of the density, temperature, and properties of the plasma that fills the space between the stars. Scientists had long anticipated these measurements, as they offer unprecedented insight into the interaction between the solar wind and the interstellar medium.

Although Voyager 2 crossed the boundary into interstellar space later than its twin, Voyager 1, in 2018, its scientific instruments have been critical in expanding our understanding of this uncharted territory. For instance, Voyager 2 confirmed an unexpected increase in the density of interstellar space, as reported in 2020, a finding that suggests the boundary between the solar system and interstellar space is more complex than previously thought.

Scientific Milestones Post-Heliosphere

Since leaving the heliosphere, Voyager 2 has made several important contributions. In 1992, the probe observed a far-ultraviolet nova, V1974 Cygni, providing astronomers with crucial data about the evolution of such stellar phenomena. Although Voyager 2 attempted to observe the impacts of fragments of Comet Shoemaker-Levy 9 with Jupiter in 1994, it was unable to detect the resulting fireballs, as they were just below the spacecraft's detection limit.

In 2007, Voyager 2 passed through the termination shock—the point where the solar wind slows down and interacts with the interstellar medium. Unlike Voyager 1, which experienced this boundary farther from the Sun, Voyager 2 encountered the termination shock roughly 1 billion miles closer, revealing that the solar system is asymmetrically shaped. This discovery provided new insights into the solar wind's interaction with the interstellar magnetic field.

In 2010, a minor glitch caused by a flipped bit in Voyager 2's onboard computer momentarily disrupted the spacecraft's ability to send scientific data. Engineers at NASA's Jet Propulsion Laboratory (JPL) quickly diagnosed and corrected the issue, restoring normal operation.

Crossing into Interstellar Space

After years of anticipation, in December 2018, NASA confirmed that Voyager 2 had officially entered interstellar space on November 5 of that year. The key evidence for this milestone came from the probe's plasma spectrometer, which detected a sharp drop in the speed of solar wind particles and a corresponding increase in the density of the interstellar plasma. This moment marked the second time a human-made object had entered the space between stars, following Voyager 1's crossing in 2012.

At approximately 120 astronomical units (AU) from the Sun, Voyager 2 has traveled beyond the last major planet of our solar system, Neptune, and into the interstellar medium. It continues to transmit invaluable data about the environment beyond the influence of our Sun, offering the first glimpse of what lies beyond the solar system's outermost boundaries.

The mission has not been without technical challenges. In November 2006, a command sent to Voyager 2 was incorrectly decoded by the spacecraft's computer, resulting in the accidental activation of its magnetometer heaters. The heaters remained on for several days, causing temperatures to rise well beyond the instrument's design limits. Fortunately, engineers quickly corrected the error, and the mission continued with no lasting damage to the spacecraft.

More recently, in 2023, a programming error caused Voyager 2's high-gain antenna to misalign by two degrees, breaking its communication link with Earth. NASA's Deep Space Network detected the spacecraft's carrier signal and, in a high-power "shout" sent from the Canberra ground station, successfully commanded Voyager 2 to reorient its antenna. This quick response restored communication, demonstrating the resilience of both the spacecraft and the team managing its mission.

As Voyager 2 continues to drift farther from the Sun, its radioisotope thermoelectric generators (RTGs), which power the spacecraft, are gradually losing their ability to generate electricity. Over the years, several scientific instruments and systems have been deactivated to conserve power. The first to be shut down was the Photopolarimeter System in 1991, followed by the termination of the scan platform and ultraviolet observations in 1998. The spacecraft's digital tape recorder was deactivated in 2007, and in 2008, the Planetary Radio Astronomy Experiment was powered down.

Despite these limitations, Voyager 2 continues to operate several key instruments, including its cosmic ray and plasma spectrometers, which are essential for studying the interstellar medium. In 2021, NASA engineers implemented a software update to reroute power and keep the remaining instruments functional, ensuring that the spacecraft continues to provide scientific data well into the mid-2020s. However, by 2030, Voyager 2 is expected to lose the ability to power its instruments, and by 2036, it will drift out of range of the Deep Space Network.

As Voyager 2 continues its journey through the galaxy, its path will eventually take it toward the star Ross 248, located 10.3 light-years from Earth. In about 42,000 years, Voyager 2 will pass within 1.7 light-years of the star. If undisturbed, the spacecraft will continue its voyage, passing by the star Sirius in roughly 296,000 years at 4.3 light-years. Although the spacecraft is not heading toward any particular destination, its journey through the Milky Way will continue indefinitely.

NASA estimates that Voyager 2 will continue to transmit weak radio signals until at least the mid-2020s, more than 48 years after it was launched. Even after its instruments fall silent, Voyager 2 will continue to travel through interstellar space, carrying with it the Voyager Golden Record, a message from humanity to any potential extraterrestrial discoverers.

Both Voyager 1 and Voyager 2 carry a gold-plated audio-visual disc known as the Voyager Golden Record. This record was created under the direction of a team led by Carl Sagan and includes a wealth of information about Earth, its inhabitants, and the diversity of life and culture on the planet. The disc contains images of Earth, greetings in 55 different languages, and a selection of music spanning different cultures and eras. It also includes natural sounds, such as the calls of whales, the cry of a baby, and the sound of waves breaking on a shore. The Golden Record was designed to communicate the richness of human life and culture to any extraterrestrial civilizations that might one day encounter the spacecraft.

As Voyager 2 continues its journey, the Golden Record serves as a testament to humanity's desire to explore the cosmos and connect with the unknown, ensuring that even long after the spacecraft's systems shut down, it will carry a message from Earth into the stars.

Mir

First module of the first modular space station launched, marking the start of the orbital assembly
First consistently inhabited long-term research space station

On December 31, 1991, the United Nations accepted the dissolution of the USSR, which ended the space race.

The Mir space station, whose name translates from Russian as "peace" or "world," was a remarkable achievement in human space exploration. Operated initially by the Soviet Union and later by the Russian Federation, Mir orbited the Earth from 1986 to 2001, serving as the first modular space station in history. Its design allowed the station to be assembled piece by piece in space over a decade, from 1986 to 1996, eventually becoming the largest artificial satellite in orbit until it was succeeded by the International Space Station (ISS).

Mir marked the peak of Soviet space technology at the time of its construction. The station provided a unique platform for long-term scientific research in a microgravity environment, where astronauts conducted experiments across various disciplines, including biology, human physiology, physics, astronomy, and meteorology. These experiments were crucial in the development of technologies necessary for the future permanent human occupation of space.

Mir was notable for being the first continuously inhabited space station, holding the record for the longest continuous human presence in space at 3,644 days—a milestone that would only be surpassed by the ISS in 2010. Among its many achievements, the station hosted cosmonaut Valeri Polyakov's record-breaking mission, where he spent 437 days and 18 hours aboard Mir from 1994 to 1995, the longest single human spaceflight in history. Over its operational lifespan of fifteen years, Mir was occupied for a total of twelve and a half years, demonstrating its capacity to support a resident crew of three, with larger crews visiting for shorter periods.

The origins of Mir date back to a decree issued on February 17, 1976, which called for the design of an improved version of the Soviet Union's earlier Salyut space stations. The Salyut program had already seen four space stations launched between 1971 and 1976, laying the groundwork for more ambitious projects. Mir's design evolved from the lessons learned from Salyut, and by 1978, the plans for Mir envisioned a station with multiple docking ports, allowing for modular expansion. This configuration was finalized in 1979, with five docking ports arranged around a spherical docking node at the station's forward end, setting the stage for a revolutionary approach to space station design.

The original plans called for relatively small 7.5-ton modules to be added to the station, but a shift in Soviet priorities led to a more ambitious design, incorporating 20-ton modules derived from the TKS spacecraft. By the early 1980s, work on Mir was progressing slowly, with resources diverted to other Soviet projects, including the Buran space shuttle program. However, in 1984, Soviet leadership renewed their commitment to the station, with an order from the Central Committee mandating that Mir be launched by early 1986, to coincide with the 27th Communist Party Congress.

The Mir Core Module, also known as DOS-7, was launched on February 19, 1986, aboard a Proton-K rocket from the Baikonur Cosmodrome, marking the beginning of a new era in space station design. As the heart of the Mir space station, the Core Module provided the primary living and working quarters for cosmonauts, as well as critical life support systems, attitude control mechanisms, and the station's main propulsion engines. Its design drew heavily on the hardware developed for the earlier Soviet Salyut stations, particularly Salyut 6 and Salyut 7, but introduced several key innovations that would make Mir unique.

One of the Core Module's most revolutionary features was its modular design, allowing for docking and expanding additional scientific and living modules. At the forward end of the module was a spherical node equipped with four docking ports. This node served as both an airlock and the station's primary docking hub, enabling the gradual expansion of Mir over the years. The rear port of the Core Module was reserved for docking the Kvant-1 module, which arrived in 1987 as the station's first expansion. Over the following years, four additional modules were docked: Kvant-2 in 1989, Kristall in 1990, Spektr in 1995, and Priroda in 1996. These modules greatly enhanced Mir's scientific capabilities and increased its habitable space, solidifying its role as a platform for long-duration missions.

The Mir Core Module itself was designed with the comfort of its crew in mind. Unlike earlier space stations, which often sacrificed living space for scientific equipment, Mir offered cosmonauts a more habitable environment. The module contained private sleeping cabins for two crew members, equipped with individual entertainment systems, exercise equipment, and a lavatory—an improvement that made long-term stays aboard the station more bearable. This focus on habitability set Mir apart from its predecessors, which were often cramped and filled with large scientific apparatus that hindered daily life.

Technological advancements were a hallmark of the Mir Core Module. The station featured larger solar arrays and enhanced power systems, providing up to 9-10 kilowatts of electrical power. Initially, Mir's solar arrays covered 76 square meters, but this was expanded to 98 square meters in 1987 with the arrival of a third array via the Kvant-1 module. This increase in power allowed Mir to support a growing number of scientific experiments and the needs of an expanded station.

Mir's docking systems also represented an evolution in space station technology. The station was equipped with the older Igla docking system and the newer, more precise Kurs system, ensuring smoother and more reliable dockings. This was crucial for maintaining the station's resupply and crew exchange capabilities, as both Soyuz spacecraft for human transport and Progress resupply vehicles regularly docked with the station.

In the early years of the Mir program, mission planners faced significant logistical challenges. The Soviet Union was working under immense pressure to meet the launch schedule for its new space station, Mir, but there was a shortage of available Soyuz spacecraft and additional modules for immediate deployment. A unique solution was devised to address these limitations: the Soyuz T-15 mission would service both Mir and the aging Salyut 7 space station, an unprecedented dual mission showcasing Soviet ingenuity.

On March 15, 1986, cosmonauts Leonid Kizim and Vladimir Solovyov launched aboard Soyuz T-15 and successfully docked with Mir, becoming the station's first crew. During their 51-day stay, they focused on activating the station's systems, configuring equipment, and unloading vital supplies delivered by the Progress 25 and 26 resupply missions. Their efforts brought Mir to operational status, laying the groundwork for its future role as a long-term space platform. However, their mission was far from over.

The Russian Mir Space station

On May 5, 1986, Kizim and Solovyov undocked from Mir and embarked on a one-day journey to Salyut 7. Once aboard, they spent another 51 days retrieving approximately 400 kilograms of scientific materials and equipment from the aging station, which was nearing the end of its operational life. This was a critical step, as the scientific data gathered aboard Salyut 7 held valuable insights for the Soviet space program. Meanwhile, back on Mir, the uncrewed Soyuz TM-1 had arrived to test the new Soyuz TM spacecraft design, which remained docked at the station for nine days.

Upon completing their tasks at Salyut 7, Kizim and Solovyov returned to Mir on June 26, 1986. They brought with them scientific experiments and 20 advanced instruments, including a multichannel spectrometer, to further Mir's research capabilities. For the final 20 days of their mission, they conducted Earth observations and tested the station's systems before returning to Earth on July 16, 1986. Their historic mission remains the only time in history that a crew transferred between two space stations, a testament to the versatility and strategic importance of Mir.

Mir was left unoccupied following their departure, but its second expedition marked the beginning of the station's full potential. On February 5, 1987, Soyuz TM-2 launched with cosmonauts Yuri Romanenko and Aleksandr Laveykin, signaling the start of a new era for Mir. Their mission aimed to further expand the station's capabilities, notably through the arrival of the Kvant-1 module.

Initially designed for the Salyut 7 program, the Kvant-1 module encountered technical delays and was reassigned to Mir. It carried critical instruments for X-ray and ultraviolet astrophysical observations and the station's first gyroscopes for attitude control. The module was launched on March 30, 1987, but issues plagued its initial docking attempts. On April 5, 1987, Kvant-1's onboard control system failed, leading to a second unsuccessful docking attempt.

Romanenko and Laveykin conducted an extravehicular activity (EVA) to inspect the docking port to resolve the issue. During the EVA, they discovered that a trash bag, accidentally left by a departing cargo ship, had become lodged between Kvant-1 and Mir, preventing a successful connection. Once the debris was removed, the module successfully docked with Mir on April 12, 1987, expanding the station's scientific capabilities and solidifying Mir's role as a pioneering platform for space exploration.

Throughout its operational life, the Mir Core Module remained the station's hub, providing the command center and primary living quarters for its inhabitants. It was equipped with communication systems that allowed real-time interaction with TsUP, the Soviet mission control center. The module housed vital exercise and medical equipment to maintain crew health during long-duration missions, reflecting the Soviet Union's growing expertise in space medicine.

The Mir Core Module's combination of advanced technology, improved living conditions, and modular expansion capabilities made it a cornerstone of Soviet space exploration. It enabled long-duration missions that not only pushed the boundaries of human endurance in space but also laid the groundwork for future international collaborations in space station development.

The Kvant-1 Astrophysics Module, launched on March 31, 1987, aboard a Proton-K rocket, marked the first significant expansion of the Soviet Mir space station. Dedicated to astrophysical research, Kvant-1 was equipped with a sophisticated array of instruments, including an X-ray telescope, ultraviolet telescope, and high-energy X-ray detectors, which allowed cosmonauts to conduct cutting-edge studies of cosmic phenomena. The module consisted of two pressurized working compartments and an unpressurized compartment used for various experiments. In addition to its scientific role, Kvant-1 housed essential life support systems, including the Elektron oxygen generator and Vozdukh carbon dioxide scrubber, critical for sustaining long-duration missions. The module also contributed to Mir's attitude control system through six gyrodynes, making it a pivotal element in both scientific endeavors and station maintenance.

Following Kvant-1, the Kvant-2 Augmentation Module was launched on November 26, 1989, also aboard a Proton-K. This module, based on the design of the TKS spacecraft, introduced new features to improve crew functionality and expand research capabilities. Divided into three sections—a cargo and instrument compartment, an experiment compartment, and an airlock for extravehicular activities (EVAs)—Kvant-2 significantly advanced the station's habitability. Its innovative water recovery system allowed urine recycling into drinkable water, and a shower improved crew hygiene, crucial for long-term spaceflight. The module housed a variety of scientific instruments, such as spectrometers and X-ray sensors, as well as the Inkubator-2, which studied the development of quail eggs in microgravity. The Soviet Ikar, a manned maneuvering unit, provided cosmonauts with enhanced mobility during spacewalks, further extending their capacity for external station maintenance.

On May 31, 1990, the Kristall Technology Module became the fourth addition to Mir. Kristall was dedicated to research in materials processing and biotechnology, featuring furnaces for metallurgy experiments and the Aniur electrophoresis unit for biotechnology studies. One of Kristall's defining characteristics was its docking compartment, which contained two APAS-89 docking ports initially intended for the Soviet Buran shuttle. Although the Buran program was discontinued, these ports became instrumental in the Shuttle-Mir program, facilitating the docking of American Space Shuttles and fostering international collaboration. Kristall was central to the scientific and diplomatic efforts that later characterized the station's operational years.

The Spektr Power Module, launched on May 20, 1995, added critical new capabilities to Mir. Spektr served as the living quarters for American astronauts during the Shuttle-Mir program and housed numerous NASA-sponsored experiments, emphasizing international cooperation in space exploration. Equipped for Earth observation, Spektr featured instruments for atmospheric and surface research, and its four solar arrays provided approximately half of Mir's electrical power. Unfortunately, in June 1997, Spektr was severely damaged during a collision with the Progress M-34 cargo spacecraft. The impact caused the module to be partially depressurized, forcing it to be sealed off, rendering it uninhabitable and limiting its operational capabilities.

The addition of the Docking Module, launched aboard Space Shuttle Atlantis (STS-74) on November 15, 1995, further enhanced the Shuttle-Mir program by simplifying Shuttle docking procedures. Previously, Mir's Kristall module had to be repositioned for Shuttle dockings. The Docking Module eliminated this requirement, allowing safer and more efficient dockings, and providing an additional docking port. This improvement facilitated multiple missions in which American astronauts visited Mir, contributing to the growing collaboration between the United States and Russia in space exploration.

On April 26, 1996, the Priroda Earth Sensing Module was launched, becoming the final addition to the Mir space station. Dedicated to Earth observation, Priroda was equipped with a suite of sensors capable of capturing data in the microwave, infrared, and visible spectral regions. The module's scientific instruments, provided by 12 different nations, symbolized international cooperation in space. Priroda's remote sensing capabilities included a synthetic aperture radar, which enabled detailed mapping and environmental monitoring of Earth's surface, further expanding the scope of Mir's research.

Beyond its pressurized modules, Mir featured several unpressurized components that extended its scientific and operational capabilities. The Sofora girder, a 14-meter-long scaffolding structure attached to Kvant-1, supported the VDU thruster block for station attitude control. By placing the thruster block farther from the station's center of mass, Sofora significantly improved fuel efficiency during orientation adjustments. Another notable unpressurized element was the Rapana girder, used to mount external experiments. Additionally, the station was equipped with two Strela cranes, which played an essential role during EVAs by allowing cosmonauts to maneuver themselves and equipment across the station's exterior, thus facilitating routine maintenance and module assembly.

Mir's power system relied on a complex array of photovoltaic (PV) solar arrays, which converted sunlight into electricity to power the station's systems and charge its nickel-cadmium batteries. These batteries provided the station with power while it passed through Earth's shadow, ensuring continuous operation throughout its orbital cycle. The station's electrical system operated at 28 volts DC, meeting the diverse power demands of Mir's numerous experiments, life support systems, and operational controls.

The combination of pressurized modules, unpressurized elements, and an advanced power system made Mir a sophisticated and versatile platform for scientific research. It supported a range of experiments in physics, biology, materials science, and Earth observation, while also fostering unprecedented international collaboration. By the time of its deorbit in 2001, Mir had set a high standard for future space stations, such as the International Space Station (ISS), shaping humanity's understanding of long-term space habitation and cooperation.

The solar arrays on Mir rotated across a 180-degree arc to track the Sun, using sensors and motors that allowed the arrays to adjust for optimal energy generation. However, the station itself needed to be oriented correctly to ensure that the arrays received adequate sunlight. When Mir passed into Earth's shadow, the station's all-sky sensor automatically reoriented the arrays to the optimal angle for reacquiring sunlight upon exiting the shadow. During these periods in darkness, the station's batteries, each with a 60 Ah capacity, powered the station until solar power could be restored.

Over Mir's operational life, the station experienced frequent power shortages, primarily due to delays in launching additional solar arrays. The initial solar arrays, each 38 m^2 in area, were launched with the core module in 1986 and provided approximately 9 kW of power. In 1987, a third array was installed on the core module, delivering an additional 2 kW. In 1989, the Kvant-2 module added two 10-meter-long arrays, each capable of producing 3.5 kW. In 1990, the Kristall module arrived with two collapsible 15-meter-long solar arrays, each providing 4 kW of power. These panels were intended to be relocated to the Kvant-1 module during a spacewalk in 1991, but the relocation was not completed until 1995 due to delays.

By the mid-1990s, many of the station's original solar arrays had degraded and were providing significantly less power than initially expected. To address this issue, the Spektr module, launched in 1995, was equipped with four new solar arrays that together spanned 126 m^2 and produced 16 kW of electricity. Two additional arrays were delivered to the station aboard the Space Shuttle Atlantis during the STS-74 mission in November 1995. One of these, the Mir cooperative solar array, featured American photovoltaic cells mounted on a Russian frame. It was installed on the unoccupied mount on Kvant-1, replacing a dorsal panel on the core module that had degraded to providing only 1 kW of power. In 1997, the second panel replaced a Kristall array on Kvant-1, completing Mir's electrical system.

Maintaining Mir in orbit required regular altitude adjustments to counteract the effects of atmospheric drag. The station typically orbited at an altitude between 354 km (220 miles) and 374 km (232 miles), traveling at a speed of 27,700 km/h (17,200 mph), completing 15.7 orbits per day. Over time, drag caused the station's orbit to decay, necessitating periodic boosts to maintain its altitude. The engines of Progress resupply spacecraft usually provided these boosts, although US Space Shuttles also performed this task during the Shuttle-Mir program. Before the arrival of the Kvant-1 module, the engines on the core module could also be used for orbit adjustment.

Attitude control on Mir was managed through a combination of control moment gyroscopes (CMGs), also known as gyrodynes, and thrusters. The station had twelve CMGs, six located on the Kvant-1 module and six on the Kvant-2 module. These gyrodynes rotated at 10,000 rpm, maintaining the station's orientation without the need for thruster burns. When a change in attitude was required, the gyrodynes were disengaged, and thrusters mounted on the modules or the VDU thruster on the Sofora girder were used to achieve the new orientation. After the adjustment, the gyrodynes were reengaged to maintain the new attitude. These changes were often necessary to support various experiments, such as orienting the station for Earth observation or astronomy experiments, which required precise aiming of instruments. For stability during materials processing experiments, the station was oriented in a gravity-gradient attitude to minimize vibrations.

Mir had several communication systems to support telemetry, data transmission, and communication between the crew, mission control, and family members. The Lira antenna mounted on the core module provided the primary communication link with Earth, which could communicate directly with ground stations or use the Luch data relay satellite system. However, as the Luch system fell into disrepair in the 1990s, communication relied more heavily on Soviet tracking ships stationed around the globe, though these too became less available after the dissolution of the Soviet Union.

During spacewalks, cosmonauts used UHF radios to communicate with each other and with the station. Spacecraft approaching or departing Mir, including Soyuz, Progress, and the Space Shuttle, also used UHF radios for docking and undocking procedures. The station's TORU system allowed remote manual control of Progress spacecraft during rendezvous operations, providing a critical backup to the automatic docking systems.

Although Mir operated in orbit where Earth's gravity was still 88% of its surface value, the station experienced continuous free fall, creating a microgravity environment. This environment allowed the crew to conduct experiments that would have been impossible under normal gravity, particularly in the fields of materials science and biology. However, this perceived weightlessness was not perfect and was influenced by several factors. Atmospheric drag, mechanical vibrations from onboard systems, orbital corrections, and tidal forces all contributed to small but significant disturbances in the station's microgravity environment. The station's orientation and operations were carefully controlled for experiments requiring precise conditions to minimize these disturbances.

Mir's environmental control and life support system (ECLSS) was essential for maintaining a habitable environment aboard the station. The system controlled atmospheric pressure, oxygen levels, and the removal of waste products, ensuring that the station remained livable for its crew. Oxygen was produced by the Elektron system, which electrolyzed water to generate breathable oxygen, venting the resulting hydrogen into space. In case of a malfunction, the station carried bottled oxygen and solid fuel oxygen generators (SFOG) as backups.

Carbon dioxide was removed from the air using the Vozdukh system, and other metabolic byproducts, such as methane and ammonia, were filtered out by activated charcoal filters. Wastewater, including condensation from the air, sink water, and urine, was recycled for reuse. This closed-loop system allowed the station to maintain a stable environment for extended periods, a capability crucial for long-duration missions in space.

The atmosphere on Mir was kept at the same pressure as at sea level on Earth—101.3 kPa (14.7 psi)—to ensure crew comfort and avoid the need for specialized suits within the station. These life support systems formed the backbone of Mir's ability to support human life in the harsh environment of space, paving the way for future space stations like the ISS, which adopted similar technologies for their own operations.

The Mir space station represented a sophisticated and dynamic engineering achievement, with its power systems, orbital control, communications, and life support working in concert to maintain a habitable environment in space. Each component of the station played a crucial role in its operation and long-term success as a platform for scientific research and international collaboration.

Mir's energy needs were met primarily through photovoltaic (PV) arrays, which converted sunlight into electricity. These arrays powered the station's systems and charged nickel-cadmium batteries, which provided electricity during the station's time in Earth's shadow. Operating on a 28-volt DC electrical system, Mir was designed to meet varying power demands, depending on the specific operational needs at any given time. The system was critical for ensuring continuous operations during the station's orbital cycle.

The solar arrays on Mir were designed to rotate across a 180-degree arc to track the Sun, optimizing energy generation. Sun sensors and motors allowed the arrays to adjust accordingly, but the station itself also needed to be correctly oriented to ensure maximum sunlight exposure. When Mir passed into Earth's shadow, the station's all-sky sensor automatically reoriented the arrays to prepare for reacquiring the Sun upon exiting the shadow. During these periods, the station's nickel-cadmium batteries, with a capacity of 60 Ah each, powered the station until solar power could be restored.

Over Mir's operational life, the station often faced power shortages due to delays in launching additional solar arrays. The first solar arrays, each covering 38 square meters, were launched with the core module in 1986, providing 9 kW of power. A third array was installed on the core module in 1987, adding 2 kW. By 1989, the Kvant-2 module introduced two 10-meter-long arrays, each supplying 3.5 kW. In 1990, the Kristall module contributed two collapsible 15-meter-long arrays, each providing 4 kW. However, their planned relocation to the Kvant-1 module was not completed until 1995.

By the mid-1990s, many of Mir's original solar arrays had degraded, providing significantly less power than expected. To address this, the Spektr module, launched in 1995, was equipped with four new solar arrays spanning 126 square meters, producing 16 kW of electricity. Two additional arrays were delivered during the STS-74 Space Shuttle mission in November 1995. The Mir cooperative solar array, featuring American PV cells on a Russian frame, replaced a failing dorsal panel on the core module. In 1997, the second array replaced the Kristall array on Kvant-1, completing Mir's electrical system.

Maintaining Mir's orbit required frequent adjustments to counteract the effects of atmospheric drag. The station orbited Earth at an altitude between 354 km and 374 km, traveling at a speed of 27,700 km/h (17,200 mph) and completing 15.7 orbits per day. Atmospheric drag gradually reduced the station's altitude, necessitating periodic boosts to maintain its orbit. These boosts were usually performed by Progress resupply spacecraft, though US Space Shuttles also performed this task during the Shuttle-Mir program. Prior to the arrival of Kvant-1, the core module's engines could also perform these orbital adjustments.

Attitude control on Mir was achieved using a combination of control moment gyroscopes (CMGs), also known as gyrodynes, and thrusters. The station had twelve CMGs, six located in the Kvant-1 module and six in the Kvant-2 module. These gyrodynes rotated at 10,000 rpm, maintaining the station's orientation without the need for thrusters. When the station's orientation needed to be adjusted, the gyrodynes were disengaged, and thrusters mounted on the modules, or the VDU thruster on the Sofora girder, were used. This system allowed for precise orientation adjustments, such as those required for Earth observation or astronomical experiments, which needed the station's instruments to be aimed continuously at a target. For materials processing experiments, the station was oriented in a gravity-gradient attitude, which minimized vibrations.

Mir relied on a variety of communication systems to maintain telemetry, data transmission, and crew communications with mission control and family members. The Lira antenna, mounted on the core module, provided the primary communication link with Earth, allowing direct communication with ground stations or through the Luch data relay satellite system. However, the Luch system fell into disrepair in the 1990s, forcing Mir to rely more heavily on Soviet tracking ships positioned globally. These ships also became less available after the dissolution of the Soviet Union.

Cosmonauts conducting spacewalks communicated via UHF radios, which also enabled communication with approaching spacecraft, such as Soyuz, Progress, and the Space Shuttle. The station's TORU system allowed remote manual control of Progress spacecraft during docking operations, providing a vital backup to the automated systems.

Although Mir orbited Earth where gravity was still 88% of that at the surface, the station was in a constant state of free fall, creating a microgravity environment. This allowed the crew to perform experiments that would have been impossible under normal gravity, particularly in the fields of materials science and biology. However, this perceived weightlessness was not perfect. The microgravity environment was disturbed by several factors, including atmospheric drag, mechanical vibrations from the station's systems, orbital corrections, and tidal forces. To minimize these disturbances during sensitive experiments, the station's orientation and operations were carefully managed.

The environmental control and life support system (ECLSS) on Mir was designed to provide a stable and livable environment for the crew during long-duration missions. The system maintained atmospheric pressure, controlled oxygen levels, and removed waste products from the air and water. Oxygen was produced by the Elektron system, which electrolyzed water to generate breathable oxygen while venting the hydrogen byproduct into space. The station also carried bottled oxygen and solid fuel oxygen generators (SFOG), known as Vika, as backup systems.

Carbon dioxide was removed using the Vozdukh system, while other metabolic byproducts, such as methane from digestion and ammonia from sweat, were filtered out by activated charcoal. Wastewater, including condensation from the air and urine, was recycled for reuse, allowing the station to maintain a closed-loop system for extended periods.

The atmosphere on Mir was maintained at sea-level pressure, 101.3 kPa (14.7 psi), ensuring the crew's comfort without the need for specialized suits inside the station. These life support systems were essential for the station's ability to support human life in space, laying the groundwork for similar systems used on the International Space Station (ISS).

Through its advanced systems of solar arrays, orbit control, communications, and life support, Mir became a model for future space stations. The station not only demonstrated the viability of long-term human habitation in space but also provided invaluable insights into the challenges of sustaining life and conducting microgravity research. Its legacy continues to influence space exploration and the development of orbital habitats today.

Life aboard Mir was a unique experience, blending the routine of daily life with the extraordinary challenges of living in space. By 1996, Mir had grown into a sprawling space station, resembling a cramped labyrinth of scientific instruments, hoses, cables, and personal items—photos, children's drawings, books, and even a guitar. Weighing 130 tonnes (140 short tons), Mir commonly housed three crew members but could support up to six for brief periods, typically around a month. Though initially designed to operate for five years, the station remained in orbit for fifteen, a testament to its durability and the skill of the international crews who maintained it.

By the time NASA astronaut John Blaha arrived on Mir, the station had been in continuous use for over a decade. Blaha noted that, apart from the more recently added Priroda and Spektr modules, Mir looked well-worn, a reflection of the many years it had been inhabited without the chance for a thorough cleaning. The station bore the marks of a long-term outpost, lived in by numerous crews from multiple nations.

The crew followed a regimented schedule, based on Moscow Time (UTC+03). With 16 sunrises and sunsets every day, the station's windows were often covered during designated night hours to simulate darkness. Each day began with an 8:00 a.m. wake-up call, followed by two hours for personal hygiene and breakfast. Work began at 10:00 a.m., and by 1:00 p.m., the crew would pause for an hour of exercise and a lunch break. The afternoon resumed with three more hours of work and an additional hour of exercise. By 7:00 p.m., the crew would prepare for dinner and the evening was theirs to spend as they wished.

During their free time, cosmonauts and astronauts found ways to relax in the confined quarters. They could observe Earth, catch up on correspondence from home, or stamp letters and drawings as proof they had flown aboard Mir. The station was equipped with books, films, and even ham radios, allowing crew members to communicate with amateur radio operators on Earth using the call signs U1MIR and U2MIR.

While the crew largely adhered to schedules meticulously planned by ground control, some astronauts like NASA's Jerry Linenger found ways to adapt. Linenger observed that the strict timetabling did not always make sense from a practical perspective and often led to fatigue and stress. In response, he adjusted his schedule to optimize efficiency and lessen mental strain. While his Russian comrades on board followed their tasks exactly as outlined, Linenger noticed that their adherence to rigid routines sometimes resulted in heightened stress levels. Nonetheless, he admired their professionalism, noting that they completed their duties with precision and dedication.

Shannon Lucid, another NASA astronaut who set the record for the longest stay in space by a woman during her time aboard Mir, compared the experience to working at an isolated Antarctic outpost. "The big difference," Lucid said, "is the isolation. You really are on your own." Lucid's reflections underscored the reality of life on Mir—while supported by ground teams, the crew was largely autonomous, dealing with the challenges of space life in real-time, far from Earth.

Exercise was critical for maintaining the health of astronauts and cosmonauts aboard Mir, where long-term exposure to weightlessness could lead to serious physiological effects. Shannon Lucid, during her stay on Mir, regularly used the station's exercise equipment to combat the adverse impacts of life in microgravity. Among the most significant health risks were muscle atrophy and the deterioration of bone density, known as spaceflight osteopenia. Additional effects included fluid redistribution, cardiovascular slowing, decreased red blood cell production, balance disorders, and a weakening of the immune system. Lesser symptoms included body mass loss, nasal congestion, sleep disturbances, facial puffiness, and increased flatulence. These conditions would generally begin to reverse upon return to Earth, but without intervention, they could cause long-term harm.

To mitigate these risks, Mir was outfitted with two treadmills—one in the core module and another in Kvant-2—as well as a stationary bicycle located in the core module. Each crew member was required to run approximately 5 kilometers (3.1 miles) and cycle the equivalent of 10 kilometers (6.2 miles) daily. Bungee cords strapped the astronauts to the treadmill to simulate weight-bearing exercise. This physical activity was considered vital in reducing the loss of muscle and bone density caused by prolonged exposure to low-gravity conditions.

Maintaining hygiene aboard Mir presented unique challenges in the weightless environment of space. The station had two space toilets, or ASUs, located in the core module and Kvant-2. These toilets used a fan-driven suction system to manage waste, similar to the system on the Space Shuttle. Cosmonauts and astronauts would fasten themselves to the toilet seat using spring-loaded restraining bars to ensure a proper seal. When the fan activated, it opened a suction hole that carried waste away through an air stream. Solid waste was collected in individual bags stored in aluminum containers, which were eventually transferred to the Progress spacecraft for disposal. Liquid waste was evacuated using a hose fitted with anatomically appropriate adapters, enabling both men and women to use the same toilet. The liquid was either recycled into drinking water through the Water Recovery System or used to produce oxygen via the Elektron system.

Aboard Mir, hygiene facilities included a shower known as the "Bania" in Kvant-2, a notable improvement from previous Soviet space stations. However, the setup time, complexity of use, and maintenance made the shower impractical. Eventually, it was converted into a steam room and later repurposed as storage space after its plumbing was removed. In its absence, crew members used wet wipes, rinse-less shampoo, and edible toothpaste, all designed to conserve water. They also used a washbasin with a plastic hood for basic cleaning tasks.

In 1998, a study revealed the growth of bacteria and larger organisms in water globules formed from moisture that had condensed behind service panels, a reminder of the ongoing challenges in maintaining a clean environment aboard the aging station.

Sleeping arrangements on Mir were both functional and confined. Two permanent crew quarters, known as Kayutkas, were located toward the rear of the core module. These phone booth-sized compartments featured tethered sleeping bags, fold-out desks, portholes, and storage for personal belongings. Visiting crew members did not have dedicated sleep modules. Instead, they attached their sleeping bags to any available space, often in Spektr before it was damaged during a collision with a Progress spacecraft that led to depressurization.

Ventilation was critical in these sleeping areas; without proper airflow, astronauts risked oxygen deprivation, as exhaled carbon dioxide could accumulate in bubbles around their heads while they slept.

Food on Mir was carefully prepared and designed to meet the nutritional needs of the crew. Meals were planned before each mission with the assistance of a dietitian, providing around 100 grams of protein, 130 grams of fat, and 330 grams of carbohydrates daily, along with necessary vitamins and minerals. Most of the food was frozen, refrigerated, or canned. One popular meal, jellied beef tongue, was heated in a small warming niche on the station's core module table. Crews typically drank tea, coffee, and fruit juices. Unlike the International Space Station, Mir also stocked small amounts of cognac and vodka, reserved for special occasions.

Over its years of operation, Mir faced increasing challenges from microbiological contamination. In the early 1990s, researchers discovered 90 species of microorganisms inside the station, and by its decommissioning in 2001, that number had risen to 140. As the station aged, mold and fungi grew behind service panels and inside air-conditioning equipment. These microorganisms, particularly mold, posed significant risks as they produced acids that could degrade the station's metal, glass, and rubber components. The foul odor caused by these molds became one of the most memorable impressions for visiting astronauts.

Researchers monitoring the International Space Station (ISS) in later years also noted the presence of microorganisms, such as strains of Enterobacter bugandensis, that while not harmful to humans, required ongoing vigilance to maintain a healthy environment in space. There were concerns that some of these microorganisms could evolve in the isolated environment of space and pose greater risks upon reentry to Earth.

Mir was visited by 28 long-duration or "principal" crews, each assigned an expedition number in the format EO-X. These expeditions, lasting from a few months to over a year, represented the core of Mir's operational life. The duration of each expedition varied; for example, the shortest was EO-28, which lasted 72 days, while Valeri Polyakov set a world record with his 437-day mission, launching with EO-14 and returning with EO-17. The typical length of a mission was around six months, with principal expedition crews usually consisting of two or three members. Often, astronauts launched with one expedition and returned with another, a testament to the station's continuous operations.

During the transition between one crew and the next, Mir was occasionally occupied by visiting crews, allowing the station to temporarily support up to six crew members during the week-long handover period. This flexibility was essential for maintaining station operations, as Mir's life support systems were built to accommodate larger crews for short durations.

The station was inhabited during four major periods: from 12 March to 16 July 1986 (EO-1), from 5 February 1987 to 27 April 1989 (EO-2 to EO-4), a nearly decade-long run from 5 September 1989 to 28 August 1999 (EO-5 to EO-27), and finally from 4 April to 16 June 2000 (EO-28). By the time of its decommissioning in 2001, Mir had hosted 104 people from 12 nations, making it the most visited spacecraft in history—a record later surpassed by the International Space Station (ISS).

The launch of Soyuz TM-2 and the successful docking of Kvant-1 marked the beginning of an intense period of station activity, which included six Soyuz launches and the hosting of three long-duration crews between 1987 and 1989. This period also saw the first international visitors to Mir. Muhammed Faris from Syria, Abdul Ahad Mohmand from Afghanistan, and Jean-Loup Chrétien from France were among the early international astronauts to visit the station, highlighting Mir's role as a platform for global collaboration in space.

Despite the growing international presence, the station faced periods of unoccupancy. After the departure of the EO-4 crew aboard Soyuz TM-7 on 27 April 1989, Mir was left without a crew until the next long-duration mission. This temporary unoccupancy highlighted the challenges of maintaining continuous operations on the station but did little to diminish Mir's impact as a pioneering hub for space exploration. Its role in fostering international cooperation and advancing scientific research set the stage for future space endeavors, including the ISS.

The launch of Soyuz TM-8 on 5 September 1989 marked the beginning of the longest continuous human presence in space, a record held by Mir until it was surpassed by the International Space Station in 2010. This period also signaled Mir's second expansion, with the addition of the Kvant-2 and Kristall modules. Cosmonauts Alexander Viktorenko and Aleksandr Serebrov docked with Mir after the station had been in a five-month hibernation, reviving it for the next phase of its operational life. On 29 September 1989, the crew prepared the docking systems for the arrival of Kvant-2, the first of several 20-tonne modules based on the TKS spacecraft from the Almaz military program.

The Kvant-2 module faced significant delays, launching only on 26 November 1989 after a 40-day postponement due to faulty computer chips. Initial problems with the module's solar arrays and docking systems required manual intervention, but Kvant-2 was successfully docked on 6 December. This module brought essential enhancements to Mir, including a second set of control moment gyroscopes (CMGs) for attitude control, improved life support systems for recycling water, and oxygen generation systems, reducing the station's reliance on ground-based resupply. Kvant-2 also featured a large airlock with a one-meter hatch and housed the Ikar backpack unit, comparable to NASA's Manned Maneuvering Unit, which was designed to support extravehicular activities (EVAs).

On 11 February 1990, the Soyuz TM-9 mission launched EO-6 crew members Anatoly Solovyev and Aleksandr Balandin to Mir. While docking, the EO-5 crew observed loose thermal blankets on the Soyuz TM-9 capsule, potentially jeopardizing reentry. Nevertheless, the mission continued, and in June 1990, Mir saw the addition of the Kristall module. Kristall contained furnaces for crystal production in microgravity, as well as biotechnology research equipment, including a small greenhouse for plant cultivation experiments. The module also featured two Androgynous Peripheral Attach System (APAS-89) docking ports designed for the Soviet Buran spacecraft. Although Buran never docked with Mir, these ports later proved invaluable during the Shuttle-Mir program, allowing docking with U.S. Space Shuttles.

Kristall's first docking attempt on 6 June 1990 was aborted due to a thruster failure, but after a successful docking on 10 June, the module was relocated to its final position opposite Kvant-2, restoring the station's balance. The mission was extended by 10 days to allow for system activations and an EVA to repair the loose thermal blankets on Soyuz TM-9. The next expedition, EO-7, arrived aboard Soyuz TM-10 on 3 August 1990, bringing live quail for biological research. One of the quail even laid an egg during the journey, which, along with 130 kilograms of experiment results, was returned to Earth aboard Soyuz TM-9.

The EO-10 mission, launched aboard Soyuz TM-13 on 2 October 1991, was the last crew to depart Earth from the Soviet Union. During their mission, the USSR dissolved, and the crew returned to Earth in March 1992 as citizens of the newly formed Russian Federation. The collapse of the Soviet Union brought significant challenges for Mir's continued operations. The newly established Russian Federal Space Agency, Roscosmos, struggled financially, forcing the indefinite postponement of the launch of the Spektr and Priroda modules, which had been intended to complete Mir's expansion.

Despite these challenges, Mir continued its operations with the launch of Soyuz TM-14 on 17 March 1992, marking the first human spaceflight launched from an independent Kazakhstan. The mission carried the EO-11 crew to Mir and also represented the beginning of a new era of international cooperation. On 17 June 1992, Russian President Boris Yeltsin and U.S. President George H. W. Bush announced what would become the Shuttle-Mir program. This collaboration between Russia and the United States proved vital in providing financial support for Roscosmos and eventually led to the completion and launch of the delayed Spektr and Priroda modules.

The EO-12 mission, launched in July 1992, welcomed French astronaut Michel Tognini aboard Mir, continuing the station's tradition of international collaboration. The EO-13 crew, which arrived in January 1993 aboard Soyuz TM-16, played a critical role in preparing Mir for the upcoming Shuttle-Mir program. Their spacecraft was equipped with the APAS-89 docking system, allowing it to dock with Kristall and test the system that would later be used by U.S. Space Shuttles.

During this period, the station was also subjected to the difficulties caused by the post-Soviet economic crisis. Communications became intermittent as the fleet of tracking ships previously maintained by Ukraine was withdrawn from service. The rising costs of crucial components, such as the Kurs docking system, strained Russia's ability to maintain Mir and led to accidents during tests of the TORU manual docking system in 1997. Additionally, Progress resupply missions often arrived with incomplete cargoes due to economic shortages, further complicating operations aboard the station.

In July 1993, the EO-14 crew launched aboard Soyuz TM-17 despite power blackouts at the Baikonur launch site and the nearby city of Leninsk. Upon arrival, Mir was fully occupied, so the Soyuz had to station-keep for half an hour while the Progress M-18 resupply craft vacated the front port for docking. The EO-13 crew departed soon after, on 22 July, just as Mir passed through the annual Perseid meteor shower. Although several micrometeoroid impacts were recorded, an EVA conducted on 28 September confirmed that the station had sustained no serious damage.

In 1994, Mir continued to expand its international partnerships, with Russian cosmonaut Sergei Krikalev becoming the first Russian to fly aboard a U.S. Space Shuttle during STS-60 in February. The year also saw the arrival of the EO-15 crew aboard Soyuz TM-18 in January, including Valeri Polyakov, who would remain aboard Mir for a record-setting 14-month stay. His extended mission, designed to study the long-term effects of space travel on the human body, paved the way for future long-duration missions aboard the ISS.

Despite the ongoing financial and logistical challenges, Mir remained a crucial platform for space research and international cooperation throughout the 1990s, setting the stage for the collaborative efforts that would define human space exploration in the following decades.

The Shuttle–Mir Program marked a significant collaboration between the United States and Russia during the 1990s, enhancing international cooperation in space exploration. The program involved a series of missions in which the Space Shuttle docked with the Russian space station Mir, facilitating long-duration stays for U.S. astronauts and paving the way for future cooperation on the International Space Station.

The first key mission in this partnership occurred on February 3, 1995, with the launch of Space Shuttle Discovery on STS-63. Called the "near-Mir" mission, Discovery approached Mir within 37 feet (11 meters), marking the first rendezvous between a Shuttle and the Russian space station. Although docking was not yet attempted, this mission served as a critical rehearsal for future docking operations and enabled equipment testing in preparation for more complex missions to come.

Shortly after, in March 1995, the EO-18 crew, which included Norman Thagard, the first U.S. astronaut to fly aboard a Russian spacecraft, arrived at Mir aboard Soyuz TM-21. This marked the beginning of U.S. long-duration missions on Mir. During EO-18, Russia launched the Spektr module, which would serve as living and working quarters for American astronauts aboard the station. Spektr carried scientific equipment from the United States and other countries, symbolizing the growing international cooperation aboard Mir. In June 1995, Space Shuttle Atlantis launched on STS-71, successfully docking with Mir on June 29. This marked the first docking of a U.S. spacecraft with a Russian space station since the Apollo–Soyuz Test Project in 1975. Atlantis delivered the EO-19 crew to Mir and returned the EO-18 crew to Earth, marking a new phase of Shuttle–Mir operations.

The year 1996 saw the continuation of these missions, with the launch of the EO-21 crew aboard Soyuz TM-23 in February. They were soon joined by American astronaut Shannon Lucid, who arrived aboard Atlantis on STS-76. Lucid's mission set a record for the longest stay in space by an American astronaut at the time, with 188 days aboard Mir. Her mission included the first joint U.S.–Russian spacewalk on Mir, during which the crew deployed the Mir Environmental Effects Payload. Lucid also witnessed the arrival of the final Mir module, Priroda, and French astronaut Claudie Haigneré, who flew to the station aboard Soyuz TM-24 as part of the Cassiopée mission.

Lucid was replaced by John Blaha in September 1996 during the STS-79 mission. Blaha's four-month stay on Mir saw significant improvements in operational procedures for docked Shuttle missions and "hand-over" techniques between long-duration U.S. crew members. His tenure aboard the station also included two spacewalks aimed at reconfiguring the station's power grid. In January 1997, Blaha returned to Earth aboard Atlantis on STS-81, and physician Jerry Linenger took his place.

Linenger's mission was marked by both historic achievements and severe challenges. He became the first American to perform a spacewalk from a foreign space station, testing the Russian-built Orlan-M spacesuit alongside Russian cosmonaut Vasili Tsibliyev. However, the mission also saw one of the most dangerous incidents in the history of spaceflight—a fire caused by a malfunctioning oxygen generator. This fire, along with subsequent system failures and a near-collision with a Progress resupply vehicle, put the crew in serious jeopardy. Additionally, a total electrical power failure left Mir in an uncontrolled tumble through space, further complicating the mission. Despite these challenges, Linenger completed his stay and was replaced by Anglo-American astronaut Michael Foale.

Foale arrived aboard Atlantis on STS-84 in May 1997. His mission proceeded without major issues until June 25, when the Progress M-34 cargo ship collided with the solar arrays on the Spektr module during a test of the manual docking system, known as TORU. The collision caused a breach in the module, leading to depressurization. The crew acted swiftly to isolate Spektr by cutting cables and closing the hatch, preventing the need to abandon the station. Efforts to restore power to Mir included a daring intra-vehicular activity (IVA), during which commander Anatoly Solovyev and flight engineer Pavel Vinogradov entered the depressurized module to run cables and inspect the damage. Foale later conducted a spacewalk to assess the external damage to Spektr.

Despite growing concerns about astronaut safety following these incidents, NASA and the U.S. Congress decided to continue the Shuttle–Mir program. In September 1997, David Wolf arrived on Mir aboard Atlantis on STS-86. His mission included a spacewalk by Vladimir Titov and Scott Parazynski, during which they installed a cap on the docking module for future repairs to Spektr. Wolf spent 119 days aboard Mir, before being replaced by Andy Thomas during the STS-89 mission in January 1998.

Thomas' mission marked the last U.S. long-duration stay aboard Mir. He remained on the station until June 1998, when he returned to Earth aboard Atlantis on STS-91, concluding the Shuttle–Mir program. The program, despite its challenges, proved invaluable in preparing NASA for the construction and operation of the International Space Station, fostering deeper collaboration between former Cold War adversaries and expanding the horizons of human space exploration.

The final days of the Mir space station marked the end of an era in human space exploration, concluding one of the most ambitious and enduring space programs of the 20th century. Following the departure of Space Shuttle Discovery on June 8, 1998, the EO-25 crew, consisting of cosmonauts Nikolai Budarin and Talgat Musabayev, remained aboard Mir. Their mission involved conducting materials experiments and performing a comprehensive inventory of the station. However, the future of Mir was uncertain. On July 2, 1998, Yuri Koptev, the director of the

Russian space agency Roscosmos, announced that due to insufficient funding, Mir would be deorbited in June 1999, marking the end of its active service.

The EO-26 crew, comprised of Gennady Padalka, Sergei Avdeyev, and physicist Yuri Baturin, arrived on August 15, 1998, aboard Soyuz TM-28. During their stay, the crew performed two spacewalks: one inside the Spektr module to reseat power cables and another outside the station to set up scientific experiments. These were delivered by Progress M-40, which also brought a substantial amount of propellant to begin adjusting Mir's orbit in preparation for its decommissioning. Despite these preparations, the launch of the first module of the International Space Station (ISS), Zarya, on November 20, 1998, prompted discussions about keeping Mir operational beyond 1999. Nevertheless, Roscosmos confirmed that it would not allocate further funding to extend the station's life.

On February 22, 1999, the EO-27 crew, consisting of Viktor Afanasyev and French astronaut Jean-Pierre Haigneré, arrived aboard Soyuz TM-29. The crew conducted three extravehicular activities (EVAs) to retrieve experiments and deploy a prototype communications antenna on the station's Sofora truss. However, with the ISS project demanding increasing resources, Russia announced on June 1, 1999, that Mir's deorbit would be delayed by six months in the hope of securing private funding to extend the station's mission. As the prospect of finding financial support waned, the crew focused on preparing Mir for its eventual deorbit. A specialized analog computer was installed to manage the station's systems, and one by one, each module was sealed off, starting with the docking module. On August 28, 1999, the crew departed Mir in Soyuz TM-29, officially ending nearly a decade of continuous human presence aboard the station. On September 7, 1999, the station's control systems, including the gyroscopes and main computer, were shut down, leaving the uncrewed Progress M-42 cargo ship to manage the gradual decay of Mir's orbit.

In the final chapter of Mir's operational life, there were brief hopes of extending its mission through private funding. One such attempt came with the Soyuz TM-30 mission, launched on April 4, 2000, by the private company MirCorp. Cosmonauts Sergei Zalyotin and Aleksandr Kaleri spent two months aboard Mir conducting repairs aimed at demonstrating the station's continued viability. While this mission was successful, and Russia remained optimistic about Mir's future, the country's commitment to the ISS left no resources to maintain the aging space station.

The decision to deorbit Mir was finalized, and the process was conducted in three stages. The first stage involved allowing natural atmospheric drag to reduce the station's orbit to an average altitude of 220 kilometers (140 miles). Progress M1-5, a modified version of the Progress-M cargo ship carrying extra fuel, docked with Mir to assist in the deorbit maneuvers. The second stage, initiated by two engine burns of Progress M1-5 on March 23, 2001, lowered the station's orbit to between 165 and 220 kilometers (103 and 137 miles). After a brief pause, the final deorbit burn began at 05:08 UTC, with Progress M1-5's engines firing for over 22 minutes to guide Mir into its final descent.

Mir re-entered Earth's atmosphere at approximately 05:44 UTC over the South Pacific Ocean, near Fiji. The intense heat of reentry caused major disintegration of the station's structure around 05:52 UTC. By 06:00 UTC, the remaining unburned fragments of Mir plunged into the ocean, marking the official end of the station's 15-year mission. Mir's fiery descent over the South Pacific closed a significant chapter in space exploration history, one that had not only advanced scientific research but had also fostered international cooperation in space.

Mir's long-term success as a space station depended heavily on its support from visiting spacecraft, primarily the Russian Soyuz and Progress vehicles, and later the U.S. Space Shuttle. These spacecraft played crucial roles in delivering crew, supplies, and experimental equipment, as well as providing emergency escape options and returning valuable data to Earth.

Initially, Mir's core module featured two docking ports—fore and aft—that allowed Soyuz and Progress spacecraft to dock. However, after the permanent attachment of the Kvant-1 module to the aft port in 1987, docking responsibilities shifted to the rear port of Kvant-1. This setup remained in place for the rest of Mir's operational life.

Both the core module's forward port and the rear port of Kvant-1 were equipped with systems to facilitate docking, including the older Igla and newer Kurs guidance systems. The core's forward port featured only Kurs, the more advanced of the two systems, which helped guide spacecraft during automated approaches.

Soyuz spacecraft, which first docked with Mir in 1986, were critical for personnel transportation, enabling crew rotations and cargo return. Additionally, they served as a lifeboat, providing a means for the crew to escape in the event of an emergency. Over Mir's lifetime, two models of Soyuz serviced the station. Soyuz T-15, which was Igla-equipped, was the only Soyuz-T variant to visit Mir. All subsequent flights used the more modern Kurs-equipped Soyuz-TM. A total of 31 Soyuz spacecraft visited Mir between 1986 and 2000, including 30 crewed missions and one uncrewed flight.

The Progress spacecraft provided essential logistical support, delivering water, fuel, food, and experimental materials to Mir. These uncrewed cargo vehicles, unlike Soyuz, were not designed to survive reentry. Once their cargo was unloaded, Progress ships were refilled with waste and deorbited, burning up in the atmosphere. To return scientific experiments, ten Progress flights carried Raduga reentry capsules, which could return around 150 kg of data and samples to Earth. Three variants of the Progress spacecraft visited Mir: the original 7K-TG model (18 flights), the Progress-M (43 flights), and the later Progress-M1 (3 flights), totaling 64 resupply missions over 14 years. Most Progress vehicles docked autonomously, but cosmonauts could use the TORU manual docking system in case of problems. This system was successfully used in most situations, except for the infamous collision of Progress M-34 in 1997, which damaged the Spektr module and caused depressurization of the station.

While Soyuz and Progress were Mir's primary lifelines, the station was also expected to be serviced by the Soviet Buran shuttle, designed to deliver large modules and return substantial amounts of cargo to Earth. The Kristall module carried two Androgynous Peripheral Attach System (APAS-89) docking ports, designed to accommodate Buran. These ports were initially intended for Buran and the planned Pulsar X-2 telescope, but following the cancellation of the Buran program, they were repurposed for use by U.S. Space Shuttles during the Shuttle–Mir program in the 1990s.

The Space Shuttle program added a new dimension to Mir's operations, facilitating crew rotations and cargo transfers on an unprecedented scale. Shuttle orbiters initially docked directly to Kristall, but clearance issues with Mir's solar arrays required Kristall to be relocated before each docking. To avoid this cumbersome process, a dedicated Mir Docking Module was added to Kristall, enabling shuttles to dock without repositioning the module or retracting solar arrays. These missions not only allowed for the exchange of American and Russian astronauts but also saw the largest transfers of cargo to and from the station. When docked to Mir, the combined structure of the shuttle and station created the largest spacecraft in history at the time, with a total mass of 250 tonnes (280 short tons). This temporary expansion greatly enhanced Mir's living and working space, further solidifying the station's status as a pioneering platform for international cooperation and space exploration.

The Mission Control Center in Korolyov, Russia, known as TsUP (ЦУП), played a critical role in managing Mir and its resupply missions. Located near the RKK Energia plant, TsUP could simultaneously oversee operations for up to ten spacecraft, although each control room was dedicated to a single program. Mir had its own designated control room, as did the Soyuz spacecraft and the Soviet Buran space shuttle program, which was later repurposed for the ISS. TsUP's structure closely resembled NASA's mission control in Houston, with key roles including the Flight Director, who provided policy guidance, and the Flight Shift Director, responsible for real-time decisions. Other vital roles included the Mission Deputy Shift Manager (MDSM) for control room operations, ground communications, and crew training, with the latter acting as the equivalent of NASA's capcom (capsule communicator), typically filled by someone with experience as the crew's lead trainer.

Three command and control modules were built for the Mir program. While one was launched into space, the second remained stored in Moscow as a backup for spare parts. The third was sold in 1997 to the Tommy

Bartlett Exploratory, an educational complex in Wisconsin Dells, Wisconsin. It became the centerpiece of their Space Exploration Wing, bringing a tangible piece of space history to the public.

As Mir aged, particularly during the Shuttle–Mir program in the 1990s, its systems began to deteriorate. Designed initially for a five-year mission, Mir remained in operation for 15 years, which resulted in a wide array of technical issues. Computer crashes, power losses, uncontrolled tumbles through space, and leaking pipes became common occurrences. Jerry Linenger, an American astronaut aboard Mir, noted that the station's cooling system had developed numerous tiny leaks, leading to the constant release of coolant. Upon re-entering the station after a spacewalk, Linenger described the station's air as having a pungent, chemical smell, raising concerns about the potential long-term health effects on the crew.

One of the most worrisome issues was the Elektron oxygen-generating system, which suffered frequent breakdowns. Crews increasingly relied on backup Vika solid-fuel oxygen generators (SFOG), despite the inherent risks. During the EO-23 mission handover in February 1997, a malfunction in the Vika system caused a fire aboard Mir, which burned for approximately 90 seconds (although Linenger contended it lasted closer to 14 minutes). The fire filled the station with thick smoke, forcing the crew to don respirators, though some were initially faulty. Several fire extinguishers mounted in newer modules were also immovable, further complicating the crew's response.

Mir faced multiple accidents, including a glancing collision between Kristall and Soyuz TM-17 during proximity operations in 1994. However, the most alarming incidents occurred during EO-23 in 1997. Besides the Vika fire, two docking accidents involving the TORU manual docking system posed significant threats to the station. During tests with Progress M-33 and Progress M-34, malfunctioning equipment caused both spacecraft to fail in their docking attempts. While Progress M-33 narrowly avoided the station, Progress M-34 collided with the Spektr module, puncturing its hull and causing a depressurization event that forced the crew to seal off Spektr permanently. This incident created a power crisis, as Spektr's solar arrays were responsible for a large portion of the station's electrical supply. It took weeks of work to restore power and stabilize the station.

Mir's position in low Earth orbit exposed its crews to higher levels of radiation, including cosmic rays and trapped protons from the South Atlantic Anomaly. During the EO-18 expedition, astronauts were exposed to an absorbed radiation dose of approximately 5.2 cGy, equivalent to 14.75 cSv, or 1133 µSv per day—about two years' worth of natural background radiation on Earth. Proximity to the station's hull and variations in radiation shielding between modules, such as Kvant-2 having better shielding than the core module, affected the radiation exposure levels for the crew. Extended exposure to such radiation increased the risk of cancer and could damage lymphocytes, crucial components of the immune system, thereby weakening immunity and increasing susceptibility to infections. Furthermore, cosmonauts experienced a higher incidence of cataracts due to radiation exposure.

Mir was also at risk from orbital debris and micrometeoroids, which could damage the station's pressurized modules, solar arrays, and other components. The debris ranged from entire spent rocket stages to paint flakes and coolant released from nuclear-powered satellites. Micrometeoroids also posed a significant risk to cosmonauts during spacewalks, as a strike could puncture their spacesuits and lead to depressurization. During meteor showers, crews would often sleep in their Soyuz ferries to be ready for an emergency evacuation if Mir were damaged.

Despite these challenges, Mir's mission persisted for years beyond its original design, providing invaluable insights into long-term human spaceflight and serving as a precursor to the International Space Station.

Epilogue

As the Cold War came to an end in 1991, the world entered a new era, and so too did space exploration. The fierce rivalry that had driven both the United States and the Soviet Union to remarkable achievements in space gave way to cooperation and collaboration. The space race that had once symbolized geopolitical tension between two superpowers gradually evolved into a global effort, where nations sought to unlock the mysteries of the universe together.

The most prominent symbol of this newfound cooperation was the International Space Station (ISS). Launched in 1998, the ISS became the most ambitious joint project in space history, bringing together space agencies from the United States, Russia, Europe, Japan, and Canada. For over two decades, it served as a beacon of scientific discovery and international partnership, hosting astronauts from around the world as they conducted groundbreaking research in microgravity, advancing our understanding of biology, physics, and human health in space. The ISS demonstrated that space was no longer the exclusive domain of superpower rivalry but a frontier where collaboration could drive humanity forward.

While international cooperation flourished, the end of the Cold War also ushered in new strategic concerns in space. The militarization of space, once a veiled component of Cold War strategy, became more transparent. The development of space-based weapons and defense systems, including anti-satellite technology, sparked debates about the future of space as a battleground. The establishment of the U.S. Space Command in 1985, followed decades later by the U.S. Space Force in 2019, marked a formal acknowledgment of space as a critical domain for national security. Other nations, including China and Russia, developed their own space-based military capabilities, underscoring that the post-Cold War era had not erased competition in space, but had shifted its focus to security and control over the vast reaches beyond Earth.

Independent space stations, once the purview of only a few nations, became more common as technological advancements made space exploration more accessible. China, for instance, launched its Tiangong space station in the early 2020s, asserting its position as a major space power. Meanwhile, private companies like SpaceX and Blue Origin, which emerged in the early 21st century, revolutionized access to space. No longer dependent solely on government programs, civilian and commercial ventures began launching satellites, sending astronauts to space, and even planning missions to the Moon and Mars. Civilian participation in space was no longer a distant dream, but a tangible reality.

With this new wave of private and commercial involvement, a renewed interest in the Moon and Mars took hold. NASA, in collaboration with international partners and private companies, set its sights on returning humans to the Moon under the Artemis program. The goal was no longer just to visit but to establish a sustainable presence, with the Moon serving as a stepping stone for future missions to Mars. The Red Planet, once the subject of science fiction, now became the focal point of plans for human exploration, with both NASA and private companies like SpaceX proposing manned missions within the coming decades.

As the world looks toward the future of space exploration, the possibilities seem boundless. Space, once a domain of geopolitical struggle, has transformed into a new frontier of collaboration, scientific discovery, and commercial opportunity. Yet, challenges remain—ensuring the peaceful use of space, managing the increasing presence of satellites and debris in Earth's orbit, and addressing the geopolitical implications of space militarization.

The future promises to bring more groundbreaking achievements: permanent lunar bases, crewed missions to Mars, the development of space tourism, and perhaps even the settlement of other planets. Humanity's presence in space will continue to grow, driven by both government and private enterprise. In this new era, space is no longer a race between nations, but a shared journey, pushing the boundaries of what we once thought possible.

As we move forward, the lessons of the past—of both cooperation and competition—will shape our approach to the cosmos. In the post-Cold War world, the final frontier remains open, beckoning us to explore, discover, and, perhaps one day, to live among the stars.

On September 14, 2024, at 7:53 AM ET, SpaceX once again made history by conducting the first civilian spacewalk, marking a significant milestone in humanity's journey into space. This unprecedented event took place aboard SpaceX's Crew Dragon spacecraft, orbiting approximately 420 kilometers above Earth. The spacewalk was part of a privately funded mission designed to push the boundaries of civilian participation in space.

Led by a crew of non-professional astronauts, the spacewalk involved stepping outside the Crew Dragon's airlock into the vacuum of space, an achievement that had previously been reserved for highly trained government astronauts from agencies like NASA, Roscosmos, and ESA. The civilians were equipped with state-of-the-art space suits designed specifically for ease of use, safety, and mobility, enabling them to perform complex tasks while experiencing the awe-inspiring view of Earth from space.

The mission, which was closely monitored by SpaceX engineers and supported by NASA's safety protocols, demonstrated the growing role of private companies in space exploration. It also showcased the increasing accessibility of space for civilians, a dream that had seemed distant only a few decades earlier. The successful spacewalk not only highlighted SpaceX's leadership in commercial space travel but also pointed toward a future where space is no longer the exclusive domain of trained astronauts, but a place where civilians can explore, work, and one day, perhaps, live.

This landmark event was a powerful reminder of how far space exploration had come since the Cold War. No longer driven solely by geopolitical rivalries, space had become a new frontier for human achievement and commercial innovation. The first civilian spacewalk marked the beginning of a new chapter in space history, opening the door to a future where anyone, given the right preparation and resources, could become a part of humanity's journey beyond Earth.

More than 15,000 satellites have been launched since the first artificial satellite, Sputnik 1, was sent into space in 1957. This number includes scientific, communication, Earth observation, and military satellites. Roughly 7,000 to 9,000 satellites are active in orbit today. The exact number fluctuates as new satellites are launched and others are decommissioned or defunct. Thousands of inactive or defunct satellites and debris from past missions contribute to the space debris issue. Estimates suggest there are more than 3,000 inactive satellites currently orbiting the Earth. Much of the growth in satellite launches in recent years has come from mega constellations like SpaceX's Starlink and OneWeb, which are deploying thousands of satellites for global internet coverage.

Several space probe missions are currently ongoing, pushing the boundaries of our knowledge about the solar system and beyond. These missions are operated by space agencies like NASA, ESA (European Space Agency), and others, with diverse goals ranging from studying distant planets to observing the Sun and exploring interstellar space.

Launched in 1977, NASA's Voyager 1 and Voyager 2 are the longest-operating space probes. Both probes have traveled far beyond the solar system's planets. Voyager 1 entered interstellar space in 2012, and Voyager 2 followed in 2018. These probes continue to send valuable data about the heliosphere—the bubble of particles and magnetic fields surrounding our solar system—and their environment beyond it. They are still operational today, although their power sources are slowly depleting, and their instruments are gradually being turned off to conserve energy.

NASA's New Horizons mission, launched in 2006, made history with its flyby of Pluto in July 2015, providing the first close-up images of the dwarf planet and its moons. After this successful encounter, New Horizons continued its journey into the Kuiper Belt, where it flew by the Kuiper Belt object Arrokoth in 2019. The probe is still operational and is currently exploring the outer regions of the solar system, transmitting data about distant celestial bodies in the Kuiper Belt.

Launched by NASA in 2011, the Juno spacecraft is currently studying Jupiter. Since its arrival in 2016, it has been providing detailed information about the planet's atmosphere, magnetic field, and composition. Juno is particularly focused on understanding Jupiter's deep interior and its role in shaping the solar system. The mission has been extended until 2025, during which time Juno will continue to explore the planet's polar regions and its large moons, including Europa, Ganymede, and Io.

NASA's OSIRIS-REx mission was launched in 2016 to study the near-Earth asteroid Bennu. After extensively mapping Bennu's surface, OSIRIS-REx successfully collected a sample of the asteroid in 2020. The spacecraft is returning to Earth, with the sample expected to arrive in 2023. However, after delivering the sample, OSIRIS-REx will embark on an extended mission called OSIRIS-APEX to explore the asteroid Apophis.

Numerous missions to Mars are still operational today. NASA's Curiosity rover, launched in 2011, continues its exploration of Gale Crater, studying Mars' climate and geology to assess past habitability. The Perseverance rover, which landed in February 2021, is exploring Jezero Crater and actively collecting samples for future retrieval as part of NASA's Mars Sample Return mission. Perseverance also supports the Ingenuity helicopter, which conducts flights to scout the Martian terrain.

Several orbiters, including NASA's Mars Reconnaissance Orbiter (MRO), ESA's ExoMars Trace Gas Orbiter, and India's Mars Orbiter Mission (Mangalyaan), are still functioning. These missions provide valuable data on Mars' atmosphere, surface, and potential for supporting life.

NASA's Parker Solar Probe, launched in 2018, is the first spacecraft to "touch" the Sun. The probe is designed to fly through the outermost layers of the Sun's atmosphere, the corona, to study the solar wind and the Sun's magnetic fields. The mission aims to improve our understanding of solar activity and its effects on the solar system. It has already provided unprecedented data on solar phenomena, and its mission will continue until at least 2025.

Launched by NASA in 2021, Lucy is a mission to study the Trojan asteroids, which share an orbit with Jupiter. These asteroids are considered remnants from the early solar system and may hold clues about its formation and evolution. Over a 12-year mission, Lucy will visit eight asteroids, including both Trojan and main-belt asteroids.

Although primarily an observatory, NASA's James Webb Space Telescope (JWST), launched in December 2021, is also an advanced probe exploring the universe. Positioned at the L2 Lagrange point, JWST provides detailed infrared observations of distant galaxies, star-forming regions, and exoplanets. It is revolutionizing our understanding of the early universe and continues to gather data that could reshape theories about cosmic evolution.

ESA's Solar Orbiter, launched in 2020, is a joint mission with NASA aimed at studying the Sun's poles and its magnetic environment. The spacecraft is gathering high-resolution images and data to better understand the solar cycle and its influence on the solar system. Solar Orbiter's mission is expected to continue until the late 2020s, providing critical insights into solar activity and space weather.

BepiColombo is a joint mission between ESA and the Japan Aerospace Exploration Agency (JAXA), launched in 2018 to study Mercury. The spacecraft is currently en route to the planet, with its main mission set to begin in 2025. BepiColombo will study Mercury's surface, magnetosphere, and composition to help scientists understand the planet's formation and evolution.

These ongoing space probe missions continue to expand our understanding of the solar system and beyond, providing vital insights into planets, asteroids, and the interstellar environment. Each mission contributes to humanity's broader exploration of the universe, addressing fundamental questions about the origins of planets, the behavior of stars, and the potential for life beyond Earth.

Meanwhile, closer to home, the accumulation of space debris in Earth's orbit presents a growing challenge. This debris field, often called "space junk" or "orbital debris," consists of defunct satellites, spent rocket stages, fragments from collisions, and other discarded or broken objects that are no longer functional. Due to the high velocities at which these objects travel—up to 8 kilometers per second in low Earth orbit (LEO)—even small fragments pose a significant risk to active satellites, spacecraft, and astronauts aboard the International Space Station (ISS).

Sources and Types of Space Debris

Space debris comes in various sizes and forms, originating from multiple sources, and can be categorized as follows:

Defunct Satellites: Many inactive communication, weather, and scientific satellites have been left in orbit for decades after reaching the end of their operational lives. These now-useless satellites continue to orbit Earth, contributing significantly to the debris field.

Rocket Stages and Boosters: Once a spacecraft reaches orbit, its spent rocket stages are often left behind. These large, often multi-ton pieces of debris can remain in space for years or even decades, slowly descending toward Earth under the influence of atmospheric drag, unless they are actively deorbited.

Fragmentation Debris: Collisions between satellites or with debris often lead to the generation of thousands of smaller pieces. This process can exponentially increase the number of objects in orbit. Fragmentation also results from explosions of spacecraft due to leftover fuel or battery malfunctions, further compounding the problem.

Lost Tools and Equipment: During extravehicular activities (spacewalks) or robotic operations, astronauts or spacecraft sometimes lose small objects like tools or components, which can remain in orbit as potential hazards.

Microscopic Particles and Paint Flecks: Even tiny particles, such as paint flakes that chip off spacecraft surfaces, travel at such high velocities that they can cause damage upon collision with operational satellites or spacecraft. These microscopic debris particles, though small, are still part of the broader debris environment and pose risks, particularly to sensitive instruments.

Orbital Regions Affected by Space Debris

The space debris field is concentrated primarily in three distinct orbital regions, each with unique challenges:

Low Earth Orbit (LEO): Extending from approximately 200 km to 2,000 km above Earth, LEO is the most densely populated region of space. This is where the ISS operates and where many Earth observation and communication satellites are located. The high concentration of objects and their extreme velocities—typically between 7 and 8 kilometers per second—make LEO particularly hazardous for active missions.

Geostationary Orbit (GEO): Situated about 35,786 km above the equator, GEO is home to weather and communication satellites that must remain fixed relative to Earth's surface. Though this region is less crowded than LEO, debris here can persist for centuries due to the absence of significant atmospheric drag. This makes GEO particularly problematic in terms of long-term debris management and mitigation.

Medium Earth Orbit (MEO): MEO, located between LEO and GEO, hosts navigation satellite systems like GPS. Although the debris field here is less concentrated than in LEO, it still presents risks to the satellites operating at these altitudes. MEO debris can persist for extended periods due to the moderate atmospheric drag at these altitudes.

The Growing Threat of Space Debris

The accumulation of space debris poses increasingly significant risks to ongoing space activities. As the number of objects in orbit continues to grow—due in part to the rapid expansion of satellite constellations like SpaceX's Starlink—so too does the likelihood of collisions. Each new collision has the potential to create thousands of additional debris fragments, further congesting Earth's orbital regions and increasing the chances of more collisions. This feedback loop is known as the Kessler Syndrome, a theoretical scenario in which space becomes so cluttered with debris that it becomes difficult or impossible to conduct future space missions.

Mitigating the effects of space debris requires international cooperation, strict regulatory frameworks, and innovative technologies to remove or deorbit debris. Without active efforts to address the growing debris field, the continued use of Earth's orbits for satellites, space stations, and future space exploration missions may face serious challenges.

Major Contributors to the Space Debris Field and several key events drive the growth of space debris:

One of the most significant events occurred in 2009, when the defunct Russian satellite Cosmos 2251 collided with the operational U.S. Iridium 33 satellite, generating thousands of debris fragments.

Nations like China, the United States, and India have conducted Anti-Satellite (ASAT) Tests that intentionally destroy satellites, creating massive amounts of debris. For example, China's 2007 destruction of the Fengyun-1C satellite created over 3,000 pieces of trackable debris.

Satellites and rocket stages can explode due to leftover fuel or battery failures, generating smaller debris fragments that can multiply over time.

Space debris poses serious threats to operational spacecraft and human spaceflight. Even small objects traveling at orbital speeds can cause significant damage:

Debris colliding with satellites can impact functioning satellites, potentially disabling them or reducing their operational lifespan. This is especially concerning for valuable communication, weather, and navigation satellites that play essential roles in daily life.

The ISS must regularly adjust its orbit to avoid collisions with debris. Astronauts on spacewalks risk being hit by high-velocity objects, even those as small as paint chips, which can puncture space suits or damage spacecraft.

Kessler Syndrome: The worst-case scenario is the so-called Kessler Syndrome, a cascading effect where collisions between debris fragments create even more debris, exponentially increasing the density of the debris field. This could make certain orbits unusable and greatly hinder future space exploration and satellite deployment.

Mitigation and Management Efforts

International space agencies and private companies are increasingly aware of the need to manage and mitigate space debris. Various strategies are being explored:

Debris Removal Technologies: Concepts like nets, harpoons, robotic arms, and even lasers are being developed to capture or deorbit larger pieces of debris. The European Space Agency's ClearSpace-1 mission, planned for 2025, aims to demonstrate such debris removal technology.

Designing for Deorbiting: Satellites and rocket stages are being designed to automatically deorbit and burn up in Earth's atmosphere after their mission is complete, reducing the long-term accumulation of debris.

Regulation and Tracking: Agencies like the U.S. Department of Defense's Space Surveillance Network (SSN) and private organizations like LeoLabs actively track and catalog objects in orbit to provide collision avoidance warnings. International guidelines are also in place, encouraging satellite operators to deorbit defunct spacecraft within 25 years.

Active Debris Avoidance: Many operational satellites and the ISS are equipped with propulsion systems to perform collision avoidance maneuvers when a potential debris threat is identified.

The space debris field continues to grow, especially as more nations and private companies launch constellations of small satellites, such as SpaceX's Starlink and OneWeb. Managing this increasing congestion will be critical to ensuring future space operations' safety and sustainability. Without active efforts to mitigate and clean up space debris, Earth's orbits could become increasingly hazardous for both crewed and uncrewed missions.

About the Author

Thornton D. "TD" Barnes is a distinguished author, entrepreneur, and former military intelligence specialist. Born in Dalhart, Texas and raised on a ranch near Clayton, New Mexico and Dalhart, Texas, he cultivated a passion for exploration. After high school in Oklahoma, Barnes embarked on a ten-year military journey, initially serving in Korea as an intelligence specialist. While in the Army, he also specialized in missile and radar electronics, defending against Soviet threats and later attending the Artillery Officer Candidate School. An injury ended his military career, but Barnes soon transitioned to aerospace endeavors. He worked on significant projects at NASA's High Range in Nevada, including the X-15, the NASA NERVA nuclear rocket project, and atomic bomb testing at the Nevada Test Site. Furthermore, he participated in the CIA's Mach 3 A-12 Project OXCART and stealth projects at Area 51.

Barnes founded and led an oil and gas exploration company outside the aerospace sphere for over 40 years, delving into uranium and gold mining ventures. In retirement, he's dedicated to preserving Area 51's history, serving as president of Roadrunners Internationale and the Nevada Aerospace Hall of Fame Director Emeritus. His contributions have been spotlighted in documentaries on National Geographic, the History Channel, and other major networks. Barnes has authored several books, including "The Secret Genesis of Area 51" and "The CIA Area 51 Chronicles." He currently resides in Henderson, Nevada, continuing to influence aerospace, exploration, and literature, focusing on the formally highly classified of the CIA's era at Area 51.

Bibliography

NASA Archives

"Korolev and Freedom of Space: 14 February 1955 – 4 October 1957". NASA. Archived from the original on October 7, 2006. Retrieved February 18, 2007.

"Yuri Gagarin: Who was the first person in space?". BBC Newsround. April 12, 2021. Retrieved July 13, 2022.

Bilstein, Roger E. (1996). Stages to Saturn: A Technological History of the Apollo/Saturn Launch Vehicles. Washington: Scientific and Technical Information Branch, National Aeronautics and Space Administration. ISBN 978-0-16-048909-9.

Brugess, Colin; Kate Doolan; Bert Vis (2003). Fallen Astronauts: Heroes Who Died Reaching for the Moon. Lincoln: University of Nebraska Press. ISBN 978-0-8032-6212-6.

Dallek, Robert (2003). An Unfinished Life: John F. Kennedy, 1917-1963. Boston: Little, Brown and Company. ISBN 978-0-316-17238-7.

Freni, Pamela (2002). Space for Women: A History of Women With the Right Stuff. Santa Ana, California: Seven Locks Press. ISBN 978-1-931643-12-2.

Gainor, Chris (2001). Arrows to the Moon: Avro's Engineers and the Space Race. Burlington, Ontario: Apogee Books. ISBN 978-1-896522-83-8.

Gatland, Kenneth (1976). Manned Spacecraft, Second Revision. New York, NY, USA: Macmillan Publishing Co., Inc. pp. 100–101. ISBN 978-0-02-542820-1.

Hall, Rex; David J. Shayler (2003). Soyuz: A Universal Spacecraft. New York: Springer–Praxis Books. ISBN 978-1-85233-657-8.

Harford, James J. (1997). Korolev: How One Man Masterminded the Soviet Drive to Beat America to the Moon (1 ed.). New York: John Wiley & Sons. ISBN 978-0-471-14853-1.

Harvey, Brian (2001). Russia in Space: The Failed Frontier?. New York: Springer–Praxis Books. ISBN 978-1-85233-203-7.

Seamans, Robert C. Jr. (April 5, 1967). "Findings, Determinations And Recommendations". Report of Apollo 204 Review Board. NASA History Office. Retrieved October 7, 2007.

Siddiqi, Asif A. (2003a). Sputnik and the Soviet Space Challenge. Gainesville: University Press of Florida. ISBN 978-0-8130-2627-5.

Siddiqi, Asif A. (2003b). The Soviet Space Race with Apollo. Gainesville: University Press of Florida. ISBN 978-0-8130-2628-2.

Thompson, Neal (2004). Light This Candle: The Life & Times of Alan Shepard—America's First Spaceman. New York: Crown Publishers. ISBN 978-0-609-61001-5.

Wolfe, Tom (2001) [1979]. The Right Stuff. New York: Bantam Books. ISBN 978-0-613-91667-7.

Yeager, Chuck; Leo Janos (1985). Yeager: An Autobiography. New York: Bantam Books. ISBN 978-0-553-05093-6.

Don't miss out!

Visit the website below and you can sign up to receive emails whenever TD Barnes publishes a new book. There's no charge and no obligation.

https://books2read.com/r/B-A-YXRJB-KLTCF

BOOKS 2 READ

Connecting independent readers to independent writers.

www.ingramcontent.com/pod-product-compliance
Lightning Source LLC
Chambersburg PA
CBHW060559120726

48002CB00010B/2735